# A Handbook of Mathematical Discourse

Charles Wells

Drawings by Peter Wells

Copyright © 2003 by Charles Wells

ISBN 0-7414-1685-9

Charles Wells
Professor Emeritus of Mathematics
Case Western Reserve University
Affiliate Scholar, Oberlin College

Drawings by Peter Wells
Website for the Handbook:
http://www.cwru.edu/artsci/math/wells/pub/abouthbk.html

*Published by:*
**INFI∞ITY**
PUBLISHING.COM
*519 West Lancaster Avenue*
*Haverford, PA 19041-1413*
*Info@buybooksontheweb.com*
*www.buybooksontheweb.com*
*Toll-free (877) BUY BOOK*
*Local Phone (610) 520-2500*
*Fax (610) 519-0261*

*Printed in the United States of America*
*Printed on Recycled Paper*
*Published October 2003*

# Contents

# Preface

## Overview

This Handbook is a report on mathematical discourse. Mathematical discourse as the phrase is used here refers to what mathematicians and mathematics students say and write

- to communicate mathematical reasoning,
- to describe their own behavior when doing mathematics, and
- to describe their attitudes towards various aspects of mathematics.

The emphasis is on the discourse encountered in post-calculus mathematics courses taken by math majors and first year math graduate students in the USA. Mathematical discourse is discussed further in the Introduction.

The Handbook describes common usage in mathematical discourse. The usage is determined by citations, that is, quotations from the literature, the method used by all reputable dictionaries. The descriptions of the problems students have are drawn from the mathematics education literature and the author's own observations.

This book is a hybrid, partly a *personal testament* and partly *documentation of research*. On the one hand, it is the personal report of a long-time teacher (not a researcher in mathematics education) who has been especially concerned with the difficulties that mathematics students have passing from calculus to more advanced courses. On the other hand, it is based on objective research data, the citations.

The Handbook is also *incomplete*. It does not cover all the words, phrases and constructions in the mathematical register, and many entries need more citations. After working on the book off and on for six years, I decided essentially to stop and publish it as you see it (after lots of tidying up). One person could not hope to write a *complete* dictionary of mathematical discourse in much less than a lifetime.

The Handbook is nevertheless a substantial probe into a very large subject. The citations accumulated for this book could be the basis for a much more elaborate and professional effort by a team of mathematicians, math educators and lexicographers who together could produce a definitive dictionary of mathematical discourse. Such an effort would provide a basis for discovering the ways in which students and non-mathematicians misunderstand what mathematicians write and say. Those misunderstandings are a major (but certainly not the only) reason why so many educated and intelligent people find mathematics difficult and even perverse.

## Intended audience

The Handbook is intended for

- *Teachers of college-level mathematics*, particularly abstract mathematics at the post-calculus level, to provide some insight into some of the difficulties their students have with mathematical language.
- *Graduate students and upper-level undergraduates* who may find clarification of some of the difficulties they are having as they learn higher-level mathematics.
- *Researchers in mathematics education*, who may find observations in this text that point to possibilities for research in their field.

The Handbook assumes the mathematical knowledge of a first year graduate student in mathematics. I would encourage students with less background to read it, but occasionally they will find references to mathematical topics they do not know about. The Handbook website contains some links that may help in finding out about such topics.

## Citations

Entries are supported when possible by **citations**, that is, quotations from textbooks and articles about mathematics. This is in accordance with standard dictionary practice [Landau, 1989], pages 151ff. As in the case of most dictionaries, the citations are not included in the printed version, but reference codes are given so that they can be found online at the Handbook website.

I found more than half the citations on JSTOR, a server on the web that provides on-line access to many mathematical journals. I obtained access to JSTOR via the server at Case Western Reserve University.

# Acknowledgments

I am grateful for help from many sources:

- Case Western Reserve University, which granted the sabbatical leave during which I prepared the first version of the book, and which has continued to provide me with electronic and library services, especially JSTOR, in my retirement.
- Oberlin College, which has made me an affiliate scholar; I have made extensive use of the library privileges this status gave me.
- The many interesting discussions on the RUME mailing list and the mathedu mailing list. The website of this book provides a link to those lists.
- Helpful information and corrections from or discussions with the following people. Some of these are from letters posted on the lists just mentioned. Marcia Barr, Anne Brown, Gerard Buskes, Laurinda Brown, Christine Browning, Iben M. Christiansen, Geddes Cureton, Tommy Dreyfus, Susanna Epp, Jeffrey Farmer, Susan Gerhart, Cathy Kessel, Leslie Lamport, Dara Sandow, Eric Schedler, Annie Selden, Leon Sterling, Lou Talman, Gary Tee, Owen Thomas, Jerry Uhl, Peter Wells, Guo Qiang Zhang, and especially Atish Bagchi and Michael Barr.
- Many of my friends, colleagues and students who have (often unwittingly) served as informants or guinea pigs.

Charles Wells
Professor Emeritus, Case Western Reserve University
Affiliate Scholar, Oberlin College

# Introduction

*Note:* If a word or phrase is in this typeface then a marginal index on the same page gives the page where more information about the word or phrase can be found. A word in **boldface** indicates that the word is being introduced or defined here.

In this introduction, several phrases are used that are described in more detail in the alphabetized entries. In particular, be warned that the definitions in the Handbook are dictionary-style definitions, not mathematical definitions, and that some familiar words are used with technical meanings from logic, rhetoric or linguistics.

## Mathematical discourse

**Mathematical discourse**, as used in this book, is the written and spoken language used by mathematicians and students of mathematics for communicating about mathematics. This is "communication" in a broad sense, including not only communicating **mathematical reasoning** (definitions and **proofs**) but also communicating approaches to problem solving, typical errors, and attitudes and behaviors connected with doing mathematics.

Mathematical discourse has three components.

- The mathematical register. When communicating mathematical reasoning and facts, mathematicians speak and write in a special register of the language (only American English is considered here) suitable for communicating mathematical arguments. In this book it is called the **mathematical register**. The mathematical register uses special technical words, as well as ordinary words, phrases and grammatical constructions with special meanings that may be different

1

from their meaning in ordinary English. It is typically mixed with **expressions** from the symbolic language (below).

- The **symbolic language** of mathematics. This is arguably not a form of English, but an independent special-purpose language. It consists of the symbolic expressions and statements used in calculation and presentation of results. For example, the statement $\frac{d}{dx}\sin x = \cos x$ is a part of the symbolic language, whereas "The derivative of the sine function is the cosine function" is not part of it.

- Mathematicians' **informal jargon**. This consists of expressions such as "**conceptual proof**" and "**intuitive**". These communicate something about the process of doing mathematics, but do not themselves communicate mathematics.

The **mathematical register** and the **symbolic language** are discussed in their own entries in the alphabetical section of the book. Informal jargon is discussed further in this introduction.

## Point of view

This Handbook is grounded in the following beliefs.

***The standard interpretation***   There is a **standard interpretation** of the **mathematical register**, including the **symbolic language**, in the sense that at least most of the time most mathematicians would agree on the meaning of most statements made in the register. Students have various other interpretations of particular constructions used in the mathematical register.

- One of their tasks as students is to learn how to extract the standard interpretation from what is said and written.
- One of the tasks of instructors is to teach them how to do that.

***Value of naming behavior and attitudes*** In contrast to computer people, mathematicians rarely make up words and phrases that describe our attitudes, behavior and mistakes. Computer programmers' informal jargon has many names for both productive and unproductive behaviors and attitudes involving programming, many of them detailed in [Raymond, 1991] (see "creationism", "mung" and "thrash" for example). The mathematical community would be better off if we emulated them by greatly expanding our informal jargon in this area, particularly in connection with dysfunctional behavior and attitudes. Having a name for a phenomenon makes it more likely that you will be aware of it in situations where it might occur and it makes it easier for a teacher to tell a student what went wrong. This is discussed in [Wells, 1995].

## Descriptive and Prescriptive

Linguists distinguish between "descriptive" and "prescriptive" treatments of language. A descriptive treatment is intended to describe the language as it is used in fact, whereas a prescriptive treatment provides rules for how the author thinks it should be used. This text is mostly descriptive. It is an attempt to describe accurately the language used by American mathematicians in communicating mathematical reasoning as well as in other aspects of communicating mathematics, rather than some ideal form of the language that they should use. Occasionally I give opinions about usage; they are carefully marked as such.

Nevertheless, the Handbook is not a textbook on how to write mathematics. In particular, it misses the point of the Handbook to complain that some usage should not be included because it is wrong.

## Coverage

The words and phrases listed in the Handbook are heterogeneous. The following list describes the main types of entries in more detail.

*Technical vocabulary of mathematics:* Words and phrases in the mathematical register that name mathematical objects, relations or properties. This is not a dictionary of mathematical terminology, and most such words ("semigroup", "Hausdorff space") are *not* included. What are included are words that cause students difficulties and that occur in courses through first year graduate mathematics. *Examples:* divide, equivalence relation, function, include, positive. I have also included briefer references to words and phrases with multiple meanings.

*Logical signalers:* Words, phrases and more elaborate syntactic constructions of the mathematical register that communicate the logical structure of a mathematical argument. *Examples:* if, let, thus. These often do not have the same logical interpretation as they do in other registers of English.

*Types of prose:* Descriptions of the types of mathematical prose, with discussions of special usages concerning them. *Examples:* definitions, theorems, labeled style.

*Technical vocabulary from other disciplines:* Some technical words and phrases from rhetoric, linguistics and mathematical logic used in explaining the usage of other words in the list. These are included for completeness. *Examples:* apposition, disjunction, metaphor, noun phrase, register, universal quantifier.

Warning: The words used from other disciplines often have ordinary English meanings as well. In general, if you see a familiar word in sans serif, you probably should look it up to see what I mean by it before you flame me based on a misunderstanding of my intention! Some words for which this may be worth doing are: context, elementary, formal, iden-

4

tifier, interpretation, name, precondition, representation, symbol, term, type, variable.

**Cognitive and behavioral phenomena** Names of the phenomena connected with learning and doing mathematics. *Examples:* mental representation, malrule, reification. Much of this (but not all) is terminology from cognitive science or mathematical education community. It is my belief that many of these words should become part of mathematicians' everyday informal jargon. The entries attitudes, behaviors, and myths list phenomena for which I have not been clever enough to find or invent names.

*Note:* The use of the name "jargon" follows [Raymond, 1991] (see the discussion on pages 3–4). This is not the usual meaning in linguistics, which in our case would refer to the technical vocabulary of mathematics.

**Words mathematicians should use:** This category overlaps the preceding categories. Some of them are my own invention (and marked as such) and some come from math education and other disciplines.

**General academic words:** Phrases such as "on the one hand ... on the other hand" are familiar parts of a general academic register and are not special to mathematics. These are generally not included. However, the boundaries for what to include are certainly fuzzy, and I have erred on the side of inclusivity.

Although the entries are of different types, they are all in one list with lots of cross references. This mixed-bag sort of list is suited to the main purpose of the Handbook, to be an aid to instructors and students. The "definitive dictionary of mathematical discourse" mentioned in the Preface may very well be restricted to the mathematical register.

The Handbook does not cover the etymology of words listed herein. Schwartzman [1994] covers the etymology of many of the technical words in mathematics. In addition, the Handbook website contains further references to this topic.

# Alphabetized Entries

**a, an**   See indefinite article.

**abstract algebra**   See algebra.

**abstraction**   An **abstraction** of a concept $C$ is a concept $C'$ that includes all instances of $C$ and that is constructed by taking as axioms certain **assertions** that are true of all instances of $C$. $C$ may already be defined mathematically, in which case the abstraction is typically a legitimate **generalization** of $C$. In other cases, $C$ may be a familiar concept or **property** that has not been given a **mathematical definition**. In that case, the mathematical definition may allow instances of the abstract version of $C$ that were not originally thought of as being part of $C$.

***Example 1***   The concept of "group" is historically an abstraction of the concept of the set of all symmetries of an **object**. The group axioms are all **true** assertions about symmetries when the **binary operation** is taken to be **composition** of symmetries.

***Example 2***   The $\epsilon$-$\delta$ definition of **continuous function** is historically an abstraction of the intuitive idea that mathematicians had about functions that there was no "break" in the **output**. This abstraction became the standard definition of "continuous", but allowed functions to be called continuous that were not contemplated before the definition was introduced.

Other examples are given under **model** and in Remark 2 under **free variable**. See also the discussions under **definition**, **generalization** and **representation**.

*Citations:* (31), (270). *References:* [Dreyfus, 1992], [Thompson, 1985].

7

**abuse of notation**   A phrase used to refer to various types of **notation** that don't have **compositional** semantics. Notation is commonly called abuse of notation if it involves **suppression of parameters** or **synecdoche** (which overlap), and examples are given under those headings. Other usage is sometimes referred to as abuse of notation, for example **identifying** two structures along an isomorphism between them. *Citations:* (82), (210), (399).

> The phrase "abuse of notation" appears to me (but not to everyone) to be deprecatory or at least apologetic, but in fact some of the uses, particularly suppression of parameters, are necessary for readability. The phrase may be an imitation of a French phrase, but I don't know its history. The English word "abuse" is stronger than the French word "abus".

*Acknowledgments:* Marcia Barr.

**accented characters**   Mathematicians frequently use an **accent** to create a new **variable** from an old one, usually to denote a **mathematical object** with some specific functional relationship with the old one. The most commonly used accents are **bar**, **check**, **circumflex**, and **tilde**.

**Example 1**   Let $X$ be a subspace of a space $S$, and let $\bar{X}$ be the closure of $X$ in $S$.

*Citations:* (66), (178).

**Remark 1** · Like accents, **primes** (the symbol $'$) may be used to denote objects functionally related to the given objects, but they are also used to create new names for objects of the same type. This latter appears to be an uncommon use for accents.

**action**   See APOS.

**affirming the consequent**   The fallacy of deducing $P$ from $P \Rightarrow Q$ and $Q$. Also called the **converse error**. This is a fallacy in **mathematical reasoning**.

*Example 1*   The student knows that if a function is differentiable, then function 104 it is continuous. He concludes [ERROR] that the absolute value function is differentiable, since it is clearly continuous.

    *Citation:* (149).

**aleph**   Aleph is the first letter of the Hebrew alphabet, written א. It is the only Hebrew letter used widely in mathematics. *Citations:* (182), (183), (315), (383).

**algebra**   This word has several different meanings in the school system of the USA, and college math majors in particular may be confused by the differences.

- **High school algebra** is primarily algorithmic and concrete in nature.
- **College algebra** is the name given to a college course, perhaps remedial, covering the material covered in high school algebra.
- **Linear algebra** may be a course in matrix theory or a course in linear transformations in a more abstract setting.
- A college course for math majors called **algebra**, **abstract algebra**, or perhaps **modern algebra**, is an introduction to groups, rings, fields and perhaps modules. It is for many students the first course in abstract mathematics and may play the role of a filter course. In some departments, linear algebra plays the role of the first course in abstraction.
- **Universal algebra** is a subject math majors don't usually see until graduate school. It is the general theory of structures with $n$-ary operations subject to equations, and is quite different in character from abstract algebra.

**algorithm**   An **algorithm** is a specific set of actions that when carried out on data (input) of the allowed type will produce an output. This is the meaning in **mathematical discourse**. There are related meanings in use:

- The algorithm may be implemented as a program in a computer language. This program may itself be referred to as the algorithm.
- In texts on the subject of algorithm, the word may be given a **mathematical definition**, turning an algorithm into a **mathematical object** (compare the uses of **proof**).

**Example 1**   One might express a simpleminded algorithm for calculating a zero of a function $f(x)$ using Newton's Method by saying

"Start with a guess $x$ and calculate $x - \dfrac{f(x)}{f'(x)}$ repeatedly until $f(x)$ gets sufficiently close to 0 or the process has gone on too long."

One could spell this out in more detail this way:

1. Choose an accuracy $\epsilon$, the maximum number of iterations $N$, and a guess $s$.
2. Let $n = 0$.
3. If $|f'(s)| < \epsilon$ then stop with the message "derivative too small".
4. Replace $n$ by $n + 1$.
5. If $n > N$, then stop with the message "too many iterations".
6. Let $r = s - \dfrac{f(s)}{f'(s)}$.
7. If $|f(r)| < \epsilon$ then stop; otherwise go to step 3 with $s$ replaced by $r$.

Observe that neither description of the algorithm is in a programming language, but that the second one is precise enough that it could be translated into most programming languages quite easily. Nevertheless, it is not a program.

*Citations:* (77), (98).

**Remark 1** It is the naive concept of abstract algorithm given in the preceding examples that is referred to by the word "algorithm" as used in mathematical discourse, *except* in courses and texts on the theory of algorithms. In particular, the mathematical definitions of algorithm that have been given in the theoretical computing science literature all introduce a mass of syntactic detail that is irrelevant for understanding particular algorithms, although the precise syntax may be necessary for *proving theorems* about algorithms, such as Turing's theorem on the existence of a noncomputable function.

**Example 2** One can write a program in Pascal and another one in C to take a list with at least three entries and swap the second and third entries. There is a sense in which the two programs, although different as programs, implement the "same" abstract algorithm.

> An "algorithm" in the meaning given here appears to be a type of process as that word is used in the APOS description of mathematical understanding. Any algorithm fits their notion of process, but whether the converse is true or not is not clear.

The following statement by Pomerance [1996] (page 1482) is evidence for this view on the use of the word "algorithm": "This discrepancy was due to fewer computers being used on the project and some 'down time' while code for the final stages of the algorithm was being written." Pomerance clearly distinguishes the algorithm from the code.

**Remark 2** Another question can be raised concerning Example 2. A computer program that swaps the second and third entries of a list might do it by changing the values of pointers or alternatively by physically moving the entries. (Compare the discussion under alias). It might even use one method for some types of data (varying-length data such as strings, for example) and the other for other types (fixed-length data). Do the two methods still implement the same algorithm at some level of abstraction?

See also overloaded notation.

*Acknowledgments:* Eric Schedler, Michael Barr.

**algorithm addiction**  Many students have the **attitude** that a problem must be solved or a **proof** constructed by an **algorithm**. They become quite uncomfortable when faced with problem solutions that involve **guessing** or **conceptual** proofs that involve little or no calculation.

***Example 1***  Recently I gave a problem in my Theoretical Computer Science class that in order to solve it required finding the largest integer $n$ for which $n! < 10^9$. Most students solved it correctly, but several wrote apologies on their paper for doing it by **trial and error**. Of course, trial and error *is* a method.

***Example 2***  Students at a more advanced level may feel insecure in the case where they are faced with solving a problem for which they know there is no known feasible **algorithm**, a situation that occurs mostly in senior and graduate level classes. For example, there are no known feasible *general* algorithms for determining if two finite **groups** given by their multiplication tables are isomorphic, and there is no algorithm at all to determine if two presentations (generators and relations) give the same group. Even so, the question, "Are the dihedral group of order 8 and the quaternion group isomorphic?" is not hard. (Answer: No, they have different numbers of elements of order 2 and 4.) I have even known graduate students who reacted badly to questions like this, but none of them got through qualifiers!

See also Example 1 under **look ahead** and the examples under **conceptual**.

**alias**  The symmetry of the square illustrated by the figure below can be described in two different ways.

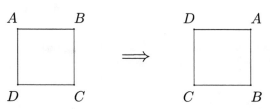

a) The corners of the square are *relabeled*, so that what was labeled $A$ is now labeled $D$. This is called the **alias** interpretation of the symmetry.

b) The square is *turned*, so that the corner labeled $A$ is now in the upper right instead of the upper left. This is the **alibi** interpretation of the symmetry.

*Reference:* These names are from [Birkhoff and Mac Lane, 1977]. They may have appeared in earlier editions of that text.

See also **permutation**.

*Acknowledgments:* Michael Barr.

**alibi**   See alias.

**all**   Used to indicate the **universal quantifier**. Examples are given under universal quantifier.

**Remark 1**   [Krantz, 1997], page 36, warns against using "all" in a **sentence** such as "All functions have a maximum", which suggests that every function has the same maximum. He suggests using **each** or **every** instead. (Other writers on mathematical writing give similar advice.) The point here is that the sentence means

$$\forall f \exists m(m \text{ is a maximum for } f)$$

not

$$\exists m \forall f(m \text{ is a maximum for } f)$$

See **order of quantifiers** and **esilism**. *Citation:* (333).

I have not found a **citation** of the form "All X have a Y" that *does* mean every $X$ has the same $Y$, and I am inclined to doubt that this is ever done. ("All" is however used to form a **collective plural** – see under collective plural for examples.) This does not mean that Krantz's advice is bad.

**always**   Used in some circumstances to indicate **universal quantification**. Unlike words such as **all** and **every**, the word "always" is attached to the verb instead of to the noun being quantified..

**Example 1**   "$x^2 + 1$ is always positive." This means, "For every $x$, $x^2 + 1$ is positive."

**Example 2**   "An ellipse always has bounded curvature."

**Remark 1**   In print, the usage is usually like Example 2, quantifying over a class of **structures**. Using "always" to quantify over a **variable** appearing in an **assertion** is not so common in writing, but it appears to me to be quite common in speech.

**Remark 2**   As the Oxford English Dictionary shows, this is a very old usage in English.

See also **never**, **time**.
*Citations:* (116), (155), (378), (424).

**ambient**   The word **ambient** is used to refer to a **mathematical object** such as a **space** that **contains** a given mathematical object. It is also commonly used to refer to an operation on the ambient space.

**Example 1**   "Let $A$ and $B$ be subspaces of a space $S$ and suppose $\phi$ is an ambient homeomorphism taking $A$ to $B$."

The point is that $A$ and $B$ are not merely homeomorphic, but they are homeomorphic via an automorphism of the space $S$.
*Citations:* (223), (172).

**analogy**   An **analogy** between two situations is a perceived similarity between some part of one and some part of the other.  Analogy, like metaphor, is a form of **conceptual blend**.

Mathematics often arises out of analogy: Problems are solved by analogy with other problems and new theories are created by analogy with older ones.  Sometimes a perceived analogy can be put in a formal setting and becomes a **theorem**.

Analogy in problem solving is discussed in [Hofstadter, 1995].

## and

**(a) *Between assertions***   The word "and" between two assertions $P$ and $Q$ produces the **conjunction** of $P$ and $Q$.

***Example 1***   The assertion

   "$x$ is **positive** and $x$ is less than 10."

is **true** if both these statements are true: $x$ is **positive**, $x$ is less than 10.

> An **argument by analogy** is the claim that because of the similarity between certain parts there must also be a similarity between some other parts. Analogy is a powerful tool that *suggests* further similarities; to use it to argue *for* further similarities is a **fallacy**.

**(b) *Between verb phrases***   The word "and" can also be used between two verb phrases to assert both of them about the same subject.

***Example 2***   The assertion of Example 1 is equivalent to the assertion

   "$x$ is positive and less than 10."

See also **both**. *Citations:* (23), (410).

**(c) *Between noun phrases***   The word "and" may occur between two noun phrases as well. In that case the translation from English statement to logical **assertion** involves subtleties.

***Example 3***   "I like red **or** white wine" means "I like red wine and I like white wine". So does "I like red and white wine". But consider also "I like red and white candy canes"!

"NO, TIMMY. THIS IS YOUR
GRANDFATHER. THE BALLS
ARE IN THE OTHER URN."

**Example 4** "John and Mary go to school" means the same thing as "John goes to school and Mary goes to school". "John and Mary own a car" (probably) does not mean "John owns a car and Mary owns a car".

On the other hand, consider also the possible meanings of "John and Mary own cars". Finally, in contrast to Examples 3 and 5, "John *or* Mary go to school" means something quite different from "John and Mary go to school."

**Example 5** In an urn filled with balls, each of a single color, "the set of red and white balls" is the same as "the set of red or white balls".

**Terminology** In mathematical logic, "and" may be denoted by "∧" or "&", or by juxtaposition.

See also the discussion under or.

**Difficulties** The preceding examples illustrate that mnemonics of the type "when you see 'and' it means intersection" *cannot work*; the translation problem requires genuine understanding of both the situation being described and the mathematical structure.

In sentences dealing with physical objects, "and" also may imply a temporal order (he lifted the weight and dropped it, he dropped the weight and lifted it), so that in contrast to the situation in mathematical assertions, "and" is not commutative in talking about physical objects. That it *is* commutative in mathematical discourse may be because mathematical objects are eternal.

As this discussion shows, to describe the relationship between English sentences involving "and" and their logical meaning is quite involved and is the main subject of [Kamp and Reyle, 1993], Section 2.4. Things are even more confusing when the sentences involve coreference, as examples in [Kamp and Reyle, 1993] illustrate.

*Acknowledgments:* The examples given above were suggested by those in the book just referenced, those in [Schweiger, 1996], and in comments by Atish Bagchi and Michael Barr.

**angle bracket** Angle brackets are the symbols "$\langle$" and "$\rangle$". They are used as outfix notation to denote various constructions, most notably an inner product as in $\langle v, w \rangle$.

*Terminology* Angle brackets are also called **pointy brackets**, particularly in speech.

*Citations:* (81), (171), (293), (105).

**anonymous notation** See structural notation.

**antecedent** The hypothesis of a conditional assertion.

**antiderivative** See integral.

**any** Used to denote the universal quantifier; examples are discussed under that heading. See also arbitrary.

**APOS** The **APOS** description of the way students learn mathematics analyzes a student's understanding of a mathematical concept as developing in four stages: **action**, **process**, **object**, **schema**.

I will describe these four ideas in terms of computing the value of a function, but the ideas are applied more generally than in that way. This discussion is oversimplified but, I believe, does convey the basic ideas in rudimentary form. The discussion draws heavily on [DeVries, 1997].

A student's understanding is at the *action stage* when she can carry out the computation of the value of a function in the following sense: after performing each step she knows how to carry out the next step.

The student is at the *process stage* when she can conceive of a process as a whole, as an algorithm, without actually carrying it out. In particular, she can describe the process step by step without having in mind a particular input.

She is at the *mathematical object* stage when she can conceive of a function as a entity in itself on which mathematical operations (for example differentiation) can be performed.

A student's **schema** for any piece of mathematics is a coherent collection of actions, processes, objects and **metaphors** that she can bring to bear on problems in that area (but see **compartmentalization**).

The APOS theory incorporates **object-process duality** and adds a stage (action) before process and another (schema) after object.

**References**  A brief overview of this theory is in [DeVries, 1997], and it is discussed in detail in [Thompson, 1994], pp. 26ff and [Asiala *et al.*, 1996], pp. 9ff. The concept of schema is similar to the concept of **procept** given in [Gray and Tall, 1994]. Schemas are discussed in the linguistic setting in [Harley, 2001], pages 329ff.

> I can attest from experience that even college students can genuinely have an understanding of a process as an action but not as a process. An attempt at teaching writing in a math course in the late sixties foundered on this when many students, most of whom could efficiently carry out long division, turned out to be incapable of writing a coherent description of the process. I did not often see students stuck at the action stage in my later years at Case Western Reserve University, when I taught mostly computer science students. They all had some programming background. Presumably that forced them to the process stage.

**arbitrary**  Used to emphasize that there is no restriction on the mathematical structure referred to by the **noun phrase** that follows. One could usually use **any** in this situation instead of "arbitrary".

**Example 1**  "The equation $x^r x^s = x^{r+s}$ holds in an arbitrary semigroup, but the equation $x^r y^r = (xy)^r$ requires commutativity." *Citations:* (249), (306), (390).

In a phrase such as "Let $S$ be an arbitrary set" the word **arbitrary** typically signals an expectation of an upcoming proof by **universal generalization**. "Any" could be used here as well.

**Difficulties**  Students are frequently bothered by constructions that seem arbitrary. Some examples are discussed under **yes it's weird**.

**argument**   This word has three common meanings in mathematical discourse.

- The input to a function may be called the argument. *Citations:* (93), (216), (429).
- The angle a complex number makes with the real axis is called the argument of the number. This can be viewed as a special case of the argument of a function.
- A proof may be called an argument. *Citations:* (7), (319).

**arity**   The **arity** of a function is the number of arguments taken by the function. The word is most commonly used for **symbols** denoting functions.

In English, "argument" can mean organized step by step reasoning to support a claim, and it can also mean the verbal expression of a disagreement. The meaning of diagreement is the common one and it carries a connotation of unpleasantness not intended by the technical meanings given here.

***Example 1***   The arity of the function sin is one.

***Example 2***   The arity of $+$ is two. It takes two arguments.

A function that takes $n$ **inputs** is also called a **function of $n$ variables**. In using the **notation** given here the order in which the **variables** are listed is important; for example, one cannot assume in general that $f(2,3) = f(3,2)$.

***Remark 1***   A function of two variables may be analyzed as a function $f : \mathbb{R} \times \mathbb{R} \to \mathbb{R}$ where $\mathbb{R} \times \mathbb{R}$ is the cartesian product of $\mathbb{R}$ with itself. In that sense it is a function with *one* input, which must be an ordered pair. I take that point of view in my class notes [Wells, 1997]; students in my class from time to time ask me why I don't write $f((x,y))$.

***Remark 2***   One sometimes finds functions with variable arity. For example, one might use MAX for the maximum of a list, and write for example

$$\mathrm{MAX}(9, 9, -2, 5) = 9$$

Of course, one might take a point of view here analogous to that of Remark 1 and say that MAX has *one* input that must be a list.

***Example 3***   Computer languages such as Lisp and Mathematica® have some functions with variable arity. The expression $+(3, 5, 5)$ in Lisp evaluates to 13, and so does the expression `Plus[3,5,5]` in Mathematica.

In general, variable arity is possible only for functions written in prefix or postfix notation *with* delimiters. When the symbol for addition (and similar symbols) is written in infix, Polish or reverse Polish notation, the symbol must have exactly two arguments. Thus the symbol $+$ in Mathematica has arity fixed at 2. *Citations:* (15), (194), (190).

*Acknowledgments:* Lou Talman.

**article**   The **articles** in English are the indefinite article "a" (with variant "an") and the definite article "the". Most of the discussion of articles is under those heads.

***Remark 1***   Both articles can cause difficulties with students whose native language does not have anything equivalent. A useful brief discussion aimed at such students is given by [Kohl, 1995]. The discussion in this Handbook is restricted to uses that cause special difficulty in mathematics.

**assertion**   An **assertion** or **statement** is a symbolic expression or English sentence (perhaps containing symbolic expressions), that may contain variate identifiers, which becomes definitely true or false if determinate identifiers are substituted for all the variate ones. If the assertion is entirely symbolic it is called a **symbolic assertion** or (in mathematical logic) a formula or sometimes, if the assertion contains variables, a predicate. Contrast with term. *Citation:* (341).

The pronunciation of a symbolic assertion may vary with its position in the discourse. See parenthetic assertion.

***Example 1***   "$2 + 2 = 4$" is an assertion. It contains no variate identifiers. In mathematical logic such an assertion may be called a sentence or proposition.

***Example 2***   "$x > 0$" is an assertion. The only variate identifier is $x$. The assertion becomes a **true** statement if 3 is substituted for $x$ and a false statement if $-3$ is substituted for $x$.

By contrast, "$x + 2y$" is not an assertion, it is a **term**; it does not become true or false when numbers are substituted for $x$ and $y$, it merely becomes an expression denoting a number.

***Example 3***   The sentence

"Either $f(x)$ is **positive** or $f(2x)$ is negative."

is an assertion. It is not a **symbolic assertion**, which in this Handbook means one that is entirely symbolic. The **variables** are $f$ and $x$ (this is discussed further under **variable**.) The assertion becomes true if cos is substituted for $f$ and $\pi/2$ is substituted for $x$. It becomes false if sin is substituted for $f$ and 0 is substituted for $x$.

***Remark 1***   It is useful to think of an assertion as a **function** with "true" and "false" as values, defined on a complicated domain consisting of statements and the possible values of their **free variables**.

*Acknowledgments:* Owen Thomas.

**assume**   See let.

**assumption**   An **assumption** is an **assertion** that is taken as an axiom in a given block of text called its **scope**. "Taken as an axiom" means that any proof in the scope of the assumption may use the assumption to justify a claim without further argument.

***Example 1***   "Throughout this chapter, $G$ will denote an arbitrary Abelian group." *Citations:* (178), (297). See also **consider**.

A statement about a physical situation may be called an assumption, as well; such statements are then taken as true for the purposes of constructing a mathematical model. *Citations:* (400), (220).

Finally, the **hypothesis** of a **conditional assertion** is sometimes called the **assumption**. *Citations:* (9), (282).

look ahead 149
trigonometric functions
  254

**at most**   For real numbers $x$ and $y$, the phrase "$x$ is at most $y$" means $x \leq y$.

*Difficulties*   Many students, including some native English speakers, do not understand this phrase. Some of them also don't understand "at least" and "not more than". *Citations:* (71), (283).

**attitudes**   Instructors, students and laymen have certain **attitudes** towards mathematics and its presentation that I think deserve names. A few are listed in this Handbook, with the page each is defined on:

| | |
|---|---|
| algorithm addiction 12 | myths 170 |
| esilism 87 | osmosis theory 188 |
| guessing 119 | Platonism 197 |
| literalist 145 | private language 204 |
| Luddism 149 | walking blindfolded 274 |
| mathematical mind 154 | yes it's weird 278 |

Here are some attitudes that need names:

**(a) *I never would have thought of that***   Example 1 under look ahead discusses the example in [Olson, 1998] of deriving a trig identity from the Pythagorean identity. One student, faced with the first step in the derivation, dividing the equation by $c$, said, "How would I ever know to divide by $c$?" I have noticed that it is common for a student to be bothered by a step that he feels he could not have thought of. My response in class to this is to say: Nevertheless, you *can* understand the proof, and *now you know a new trick.*

**(b) *No expertise required***   There seem to be subjects about which many educated people both have strong opinions and apparently are not aware that there is a body of knowledge connected with the subject. English usage is such a subject in the USA: many academicians who have

22

never read a style book and know nothing about the discoveries concerning grammar and usage that have been made in recent years nevertheless are eloquent in condemning or upholding split infinitives, commas after the penultimate entry in a series, and the like.

Happily or not, mathematics is not one of these bodies of knowledge. Non-mathematicians typically don't believe they know much about mathematics (some engineers are an exception).

On the other hand, many mathematicians take this attitude towards certain subjects; programming is one, and another is mathematics education.

*(c) I had to learn it so they should learn it*    It is noticeable that in curriculum committees professors strenuously resist relaxing a requirement that was in effect when they were students. In mathematical settings this tends to be expressed in sentences such as, "It is inconceivable that anyone could call himself a math major who has never had to **integrate** $\cos^3 x$" (or whatever). This is clearly related to the **you don't know shriek** and to **Luddism**.

*(d) I can't even balance my checkbook*    Many people who have had little association with mathematics believe that mathematics is about numbers and that mathematicians spend their time **calculating** numbers.

See also **behaviors** and **myths**.

## back formation

**back formation**    One may misread a word, perhaps derived from some root by some (often irregular) rule, as having been derived from some other nonexistent root in a more regular way. Using the nonexistent root creates a word called a **back formation**.

*Example 1*    The student who refers to a "matricee" has engaged in back formation from "matrices", which is derived irregularly from "matrix". See **plural** for more examples.

**bad at math**   See mathematical mind.

**bar**   A line drawn over a single-symbol identifier is pronounced "bar". For example, $\bar{x}$ is pronounced "x bar". Other names for this symbol are "macron" and "vinculum".

***Example 1***   "Let $F : S \to S$ be a function and $\bar{A}$ its set of fixed points."

See accent. *Citation:* (90).

*Acknowledgments:* Atish Bagchi.

**bare delimiter**   See delimiter.

**barred arrow notation**   A notation for specifying a function. It uses a barred arrow with an identifier for the input variable on the left and an expression or name that describes the value of the function on the right.

***Example 1***   "The function $x \mapsto x^2$ has exactly one critical point." Compare lambda notation and straight arrow notation. *Citation:* (273).

***Remark 1***   One can substitute input values of the correct type into barred arrow expressions, in contrast to lambda expressions (see bound variable).

***Example 2***   One can say

"Under the function $x \mapsto x^2$, one may calculate that $2 \mapsto 4$."

**be**   The verb "to be" has many uses in the English language. Here I mention a few common usages in mathematical texts.

***(a) Has a property***   For example, "The Klein four-group is Abelian." Other examples are given under property.

*(b) To define a property*  In defining a property, the word "is" may connect the **definiendum** to the name of the property, as in: "A **group** is **Abelian** if $xy = yx$ for all elements $x$ and $y$." Note that this is not an **assertion** that some group is Abelian, as in the previous entry; instead, it is saying what it means to be Abelian. *Citations:* (14), (21).

*(c) To define a type of object*  In statements such as:

> "A semigroup is a set with an associative multiplication **defined on it**."

the word "is" connects a **definiendum** with the **conditions** defining it. See **mathematical definition** for other examples. *Citations:* (40), (91), (150).

*(d) Is identical to*  The word "is" in the **statement**

> "An idempotent function has the property that its image is its set of fixed points."

asserts that two mathematical descriptions ("its image" and "its set of fixed points") denote the same **mathematical object**. This is the same as the meaning of "=". *Citations:* (27), (66), (114).

*(e) Asserting existence*  See **existential quantifier** for examples. *Citations:* (261), (346).

**behaviors**  Listed here are a number of behaviors that occur among mathematicians and students. Some of these phenomena have names (in some cases I have named them) and are discussed under that name. Many phenomena that need names are listed below. See also **attitudes** and **myths**.

*(a) Behaviors that have names*  The behaviors listed here are discussed under their names:

### (b) Behaviors that need names

*(i) All numbers are integers*  Student often unconsciously assume a number is an integer. One sometimes has scenarios in calculus classes like this:

Teacher *(with an air of triumph)*: "Now by bisection we have shown the root is between 3 and 4."

Student *(usually subvocally, but sometimes aloud)*: "But there *aren't* any numbers between 3 and 4."

*(ii) Excluding special cases*  Usually, a **generalization** of a mathematical concept will be defined in such a way as to include the **special case** it generalizes. Thus a square is a rectangle and a straight line is a curve. Students sometimes exclude the special case, saying "rectangle" to mean that the figure is *not* a square, or asking something such as "Is it a group or a semigroup?"

A definition that includes such special cases is sometimes called **inclusive**; otherwise it is **exclusive**. Most definitions in mathematics are inclusive. Exclusive definitions (such as for field or Boolean algebra) have to point out the exclusion explicitly.

*Example 1*  A field is a **nontrivial** commutative ring in which every element has an inverse. *Reference:* [Hersh, 1997a].

*(iii) Missing relational arguments*  Using a binary relation word with only one argument. For an example, see **disjoint**. Students often do this with "relatively prime".

*(iv) Forgetting to check trivial cases*

**Example 2**  A proof about **positive integers** that begins,

"Let $p$ be a prime divisor of $n$."

The integer 1 has no prime divisors.

*(v) Proving a conditional assertion backward*  When asked to prove $P \Rightarrow Q$ a student may come up with a proof beginning "If $Q \dots$ " and ending " $\dots$ therefore $P$", thus proving $Q \Rightarrow P$. This is distressingly common among students in discrete mathematics and other courses where I teach beginning **mathematical reasoning**. I suspect it comes from proving equations in high school, starting with the equation to be proved.

*(vi) Reading variable names as labels*  An assertion such as "There are six times as many students as professors" is translated by some students as $6s = p$ instead of $6p = s$ (where $p$ and $s$ have the obvious meanings). This is discussed in [Nesher and Kilpatrick, 1990], pages 101–102. See **sanity check**.

> In mathematical education, the tendency to read variable names as labels is called the **student-professor problem**, but I don't want to adopt that as a name; in some sense *every* problem in teaching is a student-professor problem!

*(vii) The representation is the object*  Many students beginning the study of abstract mathematics firmly believe that the number 735 *is* the **expression** "735". In particular, they are unwilling to use whatever **representation** of an **object** is best for the purpose.

For example, students faced with a question such as

"Does 21 divide $3 \cdot 5 \cdot 7^2$?"

will typically immediately multiply the expression out and then carry out long division to see if indeed 21 divides 735. They will say things such as,

"I can't tell what the number is until I multiply it out."
This is discussed by Brown [2002] and by Ferrari [2002].

Integers have various representations: decimal, binary, the prime factorization, and so on. Clearly the prime factorization is the best form for determining divisors, whereas for example the decimal form (in our culture) is a good form for determining which of two integers is the larger.

*(viii) Unbalanced dichotomy*    This particular incident has happened to me twice, with two different students: The students became quite upset (much more than merely puzzled) when I said, "Let $p$ be an odd prime." They were bothered because there is only one prime that is *not* odd.

The students had some expectation that is being violated, perhaps that the referents of the two parts of a dichotomy ought to be in some way balanced. Yet this example is no stranger than referring to a nonempty set. *Citation:* (416).

**binary operation**    See operation.

**black box**    See function.

**boldface**    A style of printing that **looks like this**. Section headings are often in boldface, and some authors put a definiendum in boldface. See definition.

In this text, a phrase is put in boldface in the place where it is (formally or informally) defined (except the one in the previous paragraph!).

In the symbolic language, whether a letter is in boldface or not may be significant. See also case.

***Example 1***    Let $\mathbf{v} = (v_1, v_2, v_3)$ be a vector.

*Citations:* (1), (362).

**both**   Both is an intensifier used with **and** to assert the **conjunction** of two **assertions**.

***Example 1***   "The integers 4 and 6 are both even", meaning "The integer 4 is even and the integer 6 is even". This could also be worded as: "Both the integers 4 and 6 are even." *Citations:* (119), (245), (340), (355).

***Example 2***   "2 is both even and a prime." This means "2 is even and 2 is prime." *Citation:* (251).

It is also used with **or** to emphasize that it is inclusive.

***Example 3***   "If $m$ is even and $m = rs$ then either $r$ or $s$ (or both) is even." This usage can be seen as another instance of intensifying "and": $r$ is even or $s$ is even or ($r$ is even and $s$ is even). Of course this last wording is redundant, but that is after all the point of the construction "or both". *Citations:* (75), (91).

**bound identifier**   An identifier is **bound** if it occurs in a phrase that translates directly into a **symbolic expression** in which the identifier becomes a **bound variable**. This typically occurs with the use of English quantifiers such as **all** and **every**, as well as phrases describing sums, products and integrals. An identifier that is not bound is a **free identifier**.

***Example 1***   "Any increasing function has a positive derivative." The phrase "increasing function" is bound. This **sentence** could be translated into the symbolic expression

$$\forall f \left( \text{INC}(f) \Rightarrow f' > 0 \right)$$

***Example 2***   "If an integer is even, so is its square." Here the identifier "integer" is free.

See also **variate identifier**.

**bound variable**   A variable is **bound** in a **symbolic expression** if it is within the **scope** of an **operator** that turns the **symbolic expression** into something referring collectively to all the values of the variable (perhaps within limits). The operator is said to **bind** the variable. The operators that can do this include the **existential** and **universal quantifiers**, the integral sign, the sum and product notations $\Sigma$ and $\Pi$, and various notations for **functions**. (See also **bound identifier**.) A variable that is not bound is **free**.

A key property of a bound variable is that one is not allowed to **substitute** for it (but see Example 3).

**Example 1**   In the expression $x^2 + 1$, the $x$ is a **free variable**. You can substitute 4 for $x$ in this expression and the result denotes 17. However, in $\int_3^5 x^2 + 1\, dx$, $x$ is bound by the integral sign. If you substitute 4 for $x$ you get nonsense: $\int_3^5 4^2 + 1\, d4$.

**Example 2**   In the **symbolic assertion** $x > 7$, $x$ is free. In $\forall x (x > 7)$ it is bound by the **universal quantifier** (resulting in a false statement).

**Example 3**   This example is more subtle. In the following sentence, intended to define a **function**,

"Let $f(x) = x^2 + 1$."

the **variable** $x$ is bound. It is true that one can **substitute** for the $x$ in the equation to get, for example $f(2) = 5$, but that substitution *changes the character of the statement*, from the defining equation of a function to a statement about one of its values. It is clearer that the variable $x$ is bound in this statement

"Let the function $f(x)$ be defined by $f(x) = x^2 + 1$."

which could not be transformed into

"Let the function $f(2)$ be defined by $f(2) = 2^2 + 1$."

These remarks apply also to the variables that occur in **lambda notation**, but see Example 2 under **barred arrow notation**.

***Terminology***  Bound variables are also called **dummy variables**. The latter phrase has low **status**.

***Difficulties***  Students find it difficult to learn how to use bound variables correctly.

- They may allow **variable clash**.
- They may not understand that the choice of bound variable does not matter (except for variable clash); thus $\int_2^5 x^2\,dx$ and $\int_2^5 t^2\,dt$ are the same *by their form*.
- They may move a bound variable out of its binder, for example changing $\sum_{i=1}^n i^2$ to $i\sum_{i=1}^n i$ (which makes it easy to "solve"!).
- They may **substitute** for it, although in my teaching experience that is uncommon.

***Remark 1***  The discussion in Remark 2 under **free variable** applies to bound variables as well.

***Remark 2***  Church [1942] defines "bound" as simply "not free".

## brace
**brace**  Braces are the **symbols** "{" and "}". *Citation:* (277).

A very common use of braces is in **setbuilder notation**.

***Example 1***  The set $\{(x,y) \mid y = x^2\}$ is a parabola in the plane. *Citation:* (66).

They are also used occasionally as **bare delimiters** and as **outfix notation** for functions.

***Example 2***  The expression $6/\{(1^2 + 3^2) - 2^2\}$ evaluates to 1.

***Example 3***  The fractional part of a real number $r$ is denoted by $\{r\}$. *Citations:* (265), (342), (380), (420).

A left brace may be used by itself in a definition by cases (see the example under **cases**).

***Terminology***  Braces are sometimes called **curly brackets**.

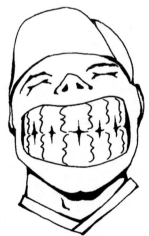

**bracket**   This word has several related usages.

**(a) *Certain* delimiters**   In common mathematical usage, **brackets** are any of the delimiters in the list

$$( ) [ ] \{ \} \langle \rangle$$

Some American dictionaries and some mathematicians restrict the meaning to **square brackets** or **angle brackets**. *Citation:* (277).

**(b) *Operation***   The word "bracket" is used in various mathematical specialties as the name of an **operation** (for example, Lie bracket, Toda bracket, Poisson bracket) in an algebra (often of operators) with a value in another structure. The operation called bracket may use square brackets, braces or angle brackets to denote the operation, but the usage for a particular operation may be fixed as one of these. Thus the Lie bracket of $v$ and $w$ is denoted by $[v, w]$. On the other hand, notation for the Poisson and Toda brackets varies. *Citations:* (415), (166), (44).

**(c) *Quantity***   The word "bracket" may be used to denote the value of the expression inside a pair of brackets (in the sense of delimiters).

***Example 1***   If the expression $(x^2 - 2x + 1) + (e^{2x} - 5)^3$ is zero, then the two brackets are opposite in **sign**. *Citation:* (414).

**but**

**(a) *And with contrast***   As a conjunction, "but" typically means the same as "and", with an indication that what follows is surprising or in contrast to what precedes it. This is a standard usage in English, not peculiar to the **mathematical register**.

***Example 1***   "5 is odd, but 6 is even." *Citations:* (111), (154), (428).

*(b) Introduces new property*  Mathematical authors may begin a sentence with "But" to indicate that the subject under discussion has a salient property that will now be mentioned, typically because it leads to the next step in the reasoning. This usage may carry with it no thought of contrast or surprise. The property may be one that is easy to deduce or one that has already been derived or assumed. Of course, in this usage "but" still means "and" as far as the logic goes; it is the connotations that are different from the usage in (a).

*Example 2*   "We have now shown that $m = pq$, where $p$ and $q$ are primes. But this implies that $m$ is composite."

*Example 3*   (In a situation where we already know that $x = 7$):
    " ... We find that $x^2 + y^2 = 100$. But $x$ is 7, so $y = \sqrt{51}$."

See also just and now.  *Reference:* [Chierchia and McConnell-Ginet, 1990], pages 283–284.

    *Citations:* (9), (123), (204).

    *Acknowledgments:* Atish Bagchi

**calculate**   To **calculate** is to perform symbol manipulation on an expression, usually with the intent to arrive at another, perhaps more satisfactory, expression.

*Example 1*   "Let us calculate the roots of the equation $x^2 - 4x + 1 = 0$."

*Example 2*   "An easy calculation shows that the equation $x^3 - 5x = 0$ factors into linear factors over the reals."

> Non-mathematicians, and many mathematics students, restrict the word "calculation" to mean arithmetic calculation, that is, symbol manipulation that comes up with a numerical answer. (See attitudes.) In contrast, I have heard mathematicians refer to calculating some object when the determination clearly involved conceptual reasoning as well as symbol manipulation.

*Example 3*   "We may calculate that $\neg\left(\forall x \exists y (x > y^2)\right)$ is equivalent to $\exists x \forall y (x \leq y^2)$."

***Remark 1***   Calculation most commonly involves arithmetic or algebraic manipulation, but the rules used may be in some other system, as Example 3 exhibits (the system is **first order logic** in that case).

See also **compute**. *Citations:* (60), (118).

**call**   Used to form a **definition**.

***Example 1***   "A monoid is called a **group** if every element has an inverse." *Citation:* (388)

***Example 2***   "Let $g = h^{-1}fh$. We call $g$ the *conjugate* of $f$ by $h$." *Citations:* (29), (101), (279).

***Example 3***   "We call an integer **even** if it is divisible by 2." *Citations:* (142), (261).

***Remark 1***   Some object to the usage in Example 3, saying "call" should be used only when you are giving a name to the object as in Examples 1 and 2. However, the usage with adjectives has been in the language for centuries. *Citation:* (267).

**cardinality**   The **cardinality** of a finite set is the number of elements of the set. This terminology is extended to **infinite** sets either by referring to the set as infinite or by using more precise words such as "countably infinite" or "uncountable".

The cardinality of a group or other **structure** is the cardinality of its underlying set.

***Difficulties***   Infinite cardinality behaves in a way that violates the expectation of students. More about this under **snow**. The book [Lakoff and Núñez, 2000] gives a deep discussion of the **metaphors** underlying cardinality and the concept of infinity in Chapters 7–10. *Citations:* (266), (409).

**case**   The Roman alphabet, the Greek alphabet, and the Cyrillic alphabet have two forms of letters, "capital" or **uppercase**, A, B, C, etc, and **lowercase**, a, b, c, etc. As far as I can tell, case distinction *always* matters in mathematics. For example, one may use a capital letter to name a **mathematical structure** and the same letter in lowercase to name an element of the structure. *Citations:* (90), (243), (317),

**Difficulties**   American students at the freshman calculus level or below quite commonly do not distinguish uppercase from lowercase when taking notes.

**Remark 1**   Other variations in font and style may also be significant. See fraktur and boldface.

**cases**   A concept is **defined by cases** if it is dependent on a **parameter** and the **definition** provides a different **defining expression** for different values of the parameter. This is also called a **disjunctive definition** or **split definition**.

**Example 1**   Let $f : \mathbb{R} \to \mathbb{R}$ be defined by

$$f(x) = \begin{cases} 1 & x > 0 \\ -1 & x \le 0 \end{cases}$$

*Citations:* (281), (384).

**Difficulties**   Students may find disjunctive definitions unnatural. This may be because real life definitions are rarely disjunctive. (One exception is the concept of "strike" in baseball.) This requires further analysis.

**category**   The word "category" is used with two unrelated meanings in mathematics (Baire and Eilenberg-Mac Lane). It is used with still other meanings by some linguists and cognitive scientists.

**character**    A **character** is a typographical symbol such as the letter "a" and the digit "3". *Citation:* (145).

     A symbol in the sense of this Handbook may consist of more than one character.

***Example 1***    The **expression** "sin" as in "$\sin \pi = 0$" is a **symbol** in the sense of this Handbook composed of three characters.

***Remark 1***    Of course, "character" also has a mathematical meaning.

**check**    The symbol "˘" over a letter is commonly pronounced "check" by mathematicians. For example, $\check{x}$ is pronounced "x check". The typographical name for this symbol is "háček". See **accent**.

**circumflex**    The symbol ˆ is a **circumflex**. Mathematicians commonly pronounce it **hat**: thus $\hat{x}$ is pronounced "x hat".

***Example 1***    "The closure of $X$ will be denoted by $\hat{X}$."

     See **accent**.

**classical category**    See **concept**.

**closed under**    A set is **closed under** an **operation** if the image of the operation is a subset of the set.

***Example 1***    The set of positive integers is closed under addition but not under subtraction. *Citation:* (206).

     *Acknowledgments:* Guo Qiang Zhang.

**code**    See **algorithm**.

**cognitive dissonance**    **Cognitive dissonance** is a term introduced in [Festinger, 1957]. It concerns conflicting understandings of some aspect of the world caused by two different modes of learning. The conflict may be resolved by suppressing the results of one of the modes of learning.

Special types of cognitive dissonance are discussed under **formal analogy**, **limit** (Example 1), **metaphor** and **semantic contamination**.

*References:* Cognitive dissonance is discussed further in [Bagchi and Wells, 1998b], [Brown and Dobson, 1996], [Carkenord and Bullington, 1993].

*Acknowledgments:* Thanks to Geddes Cureton and Laurinda Brown for suggesting references.

**collective plural**    Using the plural of an **identifier** to refer to the entire collection of items designated by the identifier.

*Example 1*    "Let $H$ be a subgroup of $G$. The left cosets of $H$ are a partition of $G$." I do not have a **citation** for this sort of wording, although I have heard people use it.

*Example 2*    "Let $\mathbb{Q}$ be the rational numbers." *Citations:* (104), (178), (356).

*Remark 1*    It appears to me that the usage shown in the two examples above is uncommon. It probably should be **deprecated**. Usually a word such as "form" or "constitute" is used, or else one refers to the set of cosets.

*Example 3*    "The left cosets of $H$ constitute a partition of $G$." or "The set of all left cosets of $H$ is a partition of $G$."

*Example 4*    "The rational numbers form a dense subset of the reals."

See **distributive plural**. *Citation:* (314).

*References:* [Lønning, 1997], [Kamp and Reyle, 1993], pages 320ff.

**college**    In the United States, a **college** is an institution one attends after graduating from high school (secondary school) that gives (usually) a B.S. or B.A. degree. A **university** also grants these degrees; the name "university" usually connotes that the institution also grants other, higher, degrees. (There are exceptions in both directions.) The usage of the word "college" is different in most other countries.

In this text, the phrase **college mathematics** denotes what in most other countries could be called "university mathematics". This is not quite correct, since in other countries much of the content of American freshman calculus would probably be taught in secondary school, or in a school that one attends between secondary school and university.

**colon equals**   The expression "$:=$" means "(is) defined to be equal to".

*Example 1*   "$S := \{1,2,3\}$ is a finite set." This is a short way of saying:

"Define $S$ to be the set $\{1,2,3\}$. Then $S$ is finite."

This usage is not very common, but my impression is that it is gaining ground.

*Remark 1*   In citations this seems to occur mostly in **parenthetic assertions**. This may be because it is hard to make an independent assertion that both is non-**redundant** and does not start with a symbol. Consider

"Let $S := \{1,2,3\}$."

(Or "Define ... ") The word "Let" already tells you we are defining $S$, so that the symbol "$:=$" is redundant. *Citations:* (28); (335). Note that although the colon equals usage is borrowed from computer languages, these two **citations** come from works in areas outside computing.

*Acknowledgments:* Gary Tee.

**combination**   An $r$-**combination** of a set $S$ is an $r$-element subset of $S$. "Combination" is the word used in combinatorics. Everywhere else in mathematics, a subset is called a subset. *Citation:* (393).

**comma**   In symbolic expressions, a comma between symbolic assertions
may denote and.

**Example 1**   The set

$$\{m \mid m = n^2,\, n \in \mathbb{Z}\}$$

denotes the set of squares of integers. The defining condition is: $m = n^2$
and $n$ is an integer.

   *Citations:* (168).

**Remark 1**   The comma is used the same way in standard written English. Consider "A large, brown bear showed up at our tent".

   The comma may also be used to indicate many-to-one coreference,

**Example 2**   "Let $x, y \neq 0$."

   *Citations:* (112), (313), (317).
   *Acknowledgments:* Michael Barr.

**compartmentalization**   A student may have several competing ways
of understanding a concept that may even be inconsistent with each other.

**Example 1**   When doing calculus homework, a student may think of
functions exclusively in terms of defining expressions, in spite of the fact
that she can repeat the ordered-pairs definition when asked and may
even be able to give an example of a function in terms of ordered pairs,
not using a defining expression. In other words, defining expressions are
for doing homework except when the question is "give the definition of
'function' "!

   This phenomenon is called **compartmentalization**. The student
has not constructed a coherent schema for (in this case) "function".

   *References:* [Tall and Vinner, 1981], [Vinner and Dreyfus, 1989].

**componentwise**   See coordinatewise.

**composite**   The **composite** of two functions $f : A \to B$ and $g : B \to C$ is a function $h : A \to C$ defined by $h(a) = g(f(a))$ for $a \in A$. It is usually denoted by $g \circ f$ or $gf$. If $A = B = C$ has a multiplicative structure, then $gf$ may also denote the function defined by $gf(a) = g(a)f(a)$, a fact that sometimes causes confusion.

When displayed using **straight arrow notation** like this

$$A \xrightarrow{\ f\ } B \xrightarrow{\ g\ } C$$

then some find the notation $g \circ f$ jarring. See **rightists**.     *Citations:* (230), (327)

The usual name for this operation is **composition**. The **result** of the operation, the function $g \circ f$, is the **composite**. See **value** for discussion of this point.

---

The operation $(g, f) \mapsto g \circ f$ is defined here only when the **codomain** of $f$ is the same as the **domain** of $g$. Many authors allow $g \circ f$ to be defined when the codomain of $f$ is included in (not necessarily equal to) the domain of $g$. Indeed, in most of the literature one cannot tell which variant of the definition is being used.

---

**Remark 1**   "Composite" has another meaning: an **integer** is composite if it has more than one prime factor.

**compositional**   The meaning of an expression is **compositional** if its meaning can be determined by the meaning of its constituent parts and by its **syntax**. Such **semantics** is also called **synthetic** or **syntax-driven**.

**Example 1**   The expression "three cats" is completely determined by the meaning of "three" and "cats" and the English rule that adjectives come before nouns. (The meaning of "cats" could of course be whips or animals; that is determined by context.)

In contrast, the meaning of a word cannot usually be synthesized from its spelling; the relationship between spelling and meaning is essentially arbitrary. As an example, consider the different roles of the letter "i" in the **symbol** "sin" and in the **expression** "$3 - 2i$".

**Remark 1** In spite of Example 1, the meaning of general English discourse is commonly *not* compositional.

***Compositionality in the symbolic language*** The symbolic language of mathematics has compositional semantics, with some exceptions. Some examples are given under **symbolic expression**; see also **syntax**.

Most of the examples of failure of compositionality that I have been able to find are examples of one of the following four phenomena (which overlap, but no one of them includes another):

- context sensitivity.
- conventions.
- suppression of parameters.
- synecdoche.

Examples are given under those headings.

**Remark 2** Some symbolic expressions are multivalued, for example

$$\int x^2\,dx$$

which is determined only up to an added constant. I don't regard this as failure of compositionality; the standard meaning of the expression is multivalued.

**compute** "Compute" is used in much the same way as **calculate**, except that it is perhaps more likely to imply that a computer was used.

**Remark 1** As in the case of **calculate**, research mathematicians often refer to computing an object when the process involves **conceptual** reasoning as well as **symbol manipulation**. *Citations:* (60), (207), (220), (331).

**concept** Mathematical concepts given by mathematical **definitions** always have the several characteristic properties:

***Equal status*** All objects that satisfy the definition have equal logical status.

***Crisp*** An object either satisfies the definition or not. An algebraic structure is either a group or it is not, but one can argue about whether Australia is a continent or a large island. Thus a mathematical object is **crisp** as opposed to **fuzzy**.

***Accumulation of attributes*** An **object** is an instance of the concept **if and only if** it has all the attributes required of it by the definition. Mathematical concepts are thus defined by an **accumulation of attributes**.

Most human concepts are not given by accumulation of attributes and many are not crisp. Furthermore, the concepts typically have internal structure, for example they may be **radial** and they may have **prototypes**. This is discussed by Lakoff in [1986], especially the discussion in Section 1, and by Pinker and Prince in [1999]. The latter reference distinguishes between **family resemblance categories** and **classical categories**; the latter are those that in my terminology are defined by accumulation of attributes.

**Remark 1** Many mathematical concepts are **abstractions** of a prior, non-mathematical concept that may be fuzzy, and one can argue about whether the mathematical definition captures the prior concept. Note also the discussions beginning on page 69 under **definition**.

***Difficulties*** Of course every student's and every mathematician's **mental representation** of a mathematical concept has more internal structure than merely the accumulation of attributes. Some instances loom large as **prototypical** and others are called by rude names such as **pathological** because they are unpleasant in some way.

Students may expect to reason with mathematical concepts using prototypes the way they (usually unconsciously) reason about everyday concepts. (See **generalization**.) On the other hand, students with some skill in handling mathematical concepts can shift psychologically between this extra internal structure and the bare structure given by accumulation of attributes, using the first for motivation and new ideas and the second in **proofs**. This shifting in the general context of human reasoning is discussed in [Pinker and Prince, 1999], section 10.4.4.

*References:* [Bagchi and Wells, 1998b], [Bagchi and Wells, 1998a], [Gopen and Smith, 1990], pages 3–6, [Vinner, 1992]. Definitions in science in general are discussed by [Halliday and Martin, 1993] pages 148–150, 170ff, 209ff.

*Acknowledgments:* To Michael Barr for pointing out sloppy thinking in a previous version of this entry, and to Tommy Dreyfus and Jeffrey Farmer for several helpful references.

> Most definitions even in science outside of mathematics are *not* by accumulation of attributes. Scientific definitions are discussed in detail in [Halliday and Martin, 1993], who clearly regard accumulation of attributes as a minor and exceptional method of definition; they mention this process in Example 13 on page 152 almost as an afterthought.

**concept image**   See mental representation.

**conceptual**   A proof is **conceptual** if it is an argument that makes use of one's **mental representation** or geometric **insight**. It is opposed to a proof by **symbol manipulation**.

**Example 1**   Let $m$ and $n$ be positive integers, and let $r$ be $m$ **mod** $n$. One can give a conceptual proof that

$$\mathrm{GCD}(m, n) = \mathrm{GCD}(m, r)$$

by showing that the set of common **divisors** of $m$ and $n$ is the same as the set of common divisors of $n$ and $r$ (easy). The result follows because the GCD of two numbers is the *greatest* common divisor, that is, the maximum of the set of common divisors of the two numbers, and a set of numbers has only one maximum.

I have shown my students this proof many times, but they almost never reproduce it on an examination.

**Example 2**   Now I will provide three proofs of a certain **assertion**, adapted from [Wells, 1995].

The statement to prove is that for all $x$, $y$ and $z$,

$$(x > z) \Rightarrow ((x > y) \lor (y > z)) \tag{1}$$

**(a) *Conceptual proof*** We may visualize $x$ and $z$ on the real line as in this picture:

There are three different regions into which we can place $y$. In the left two, $x > y$ and in the right two, $y > z$. End of proof.

This proof is written in English, not in **symbolic notation**, and it refers to a particular **mental representation** of the structure in question (the usual ordering of the real numbers).

**(b) *Symbolic Proof*** The following proof is due to David Gries (private communication) and is in the format advocated in [Gries and Schneider, 1993]. The proof is based on these principles:

P.1 (**Contrapositive**) The equivalence of $P \Rightarrow Q$ and $\neg Q \Rightarrow \neg P$.

P.2 (DeMorgan) The equivalence of $\neg(P \lor Q)$ and $\neg P \neg Q$.

P.3 The equivalence in any totally ordered set of $\neg(x > y)$ and $x \le y$.

In this proof, "$\neg$" denotes negation.

Proof:
$$(x > z) \Rightarrow ((x > y) \lor (y > z))$$
$$\equiv \quad \text{by P.1}$$
$$\neg((x > y) \lor (y > z)) \Rightarrow \neg(x > z)$$
$$\equiv \quad \text{by P.2}$$
$$(\neg(x > y)\neg(y > z)) \Rightarrow \neg(x > z)$$
$$\equiv \quad \text{by P.3 three times}$$
$$((x \le y)(y \le z)) \Rightarrow (x \le z)$$

which is true by the transitive law.

This proof involves **symbol manipulation** using logical rules and has the advantage that it is easy to check mechanically. It also shows that the proof works in a wider context (any totally ordered set).

*(c) Another conceptual proof* The conceptual proof given above provides a geometric visualization of the situation required by the **hypothesis** of the theorem, and this visualization makes the truth of the theorem obvious. But there is a sense of "conceptual", related to the idea of conceptual definition given under **elementary**, that does not have a geometric component. This is the idea that a proof is conceptual if it appeals to concepts and theorems at a high level of abstraction.

To a person familiar with the elementary rules of **first order logic**, the symbolic proof just given becomes a conceptual proof (this happened to me): "Why, in a totally ordered set that statement is nothing but the contrapositive of transitivity!" Although this statement is merely a summary of the symbolic proof, it is enough to enable anyone conversant with simple logic to generate the symbolic proof. Furthermore, in my case at least, it provides an **aha** experience. *Citations:* (360), (48).

**conceptual blend**   A cognitive structure (**concept, mental representation** or imagined situation) is a **conceptual blend** if it consists of features from two different cognitive structures, with some part of one structure merged with or identified with an isomorphic part of the other structure.

*Example 1*   An experienced mathematician may conceive of the function $x \mapsto x^2$ as represented by the parabola that is its graph, or as a machine that given $x$ produces its square (one may even have a particular **algorithm** in mind). In visualizing the parabola, she may visualize a geometric object, a curve of a certain shape placed in the plane in a certain way, and she will keep in mind that its points are parametrized (or

*Mathematics is the art of giving the same name to different things.*
*–Henri Poincaré*

45

identified with) the set $\{(x,y) \mid y = x^2\}$. The cognitive structure involved with the machine picture will include the set of paired inputs and outputs of the machine. Her complex **mental representation** of the functions includes all these objects, but in particular the pairs that parametrize the parabola and the input-output pairs of the machine are visualized as being the *same pairs*, the elements of the set $\{(x,y) \mid y = x^2\}$. That identification of the set of pairs is a conceptual blend.

**Example 2**  A monk starts at dawn at the bottom of a mountain and goes up a path to the top, arriving there at dusk. The next morning at dawn he begins to go down the path, arriving at dusk at the place he started from on the previous day. Prove that there is a time of day at which he is at the same place on the path on both days.

Proof: Envision both events occurring on the *same* day, with a monk starting at the top and another starting at the bottom and doing the same thing the monk did on different days. They are on the same path, so they must meet each other. The time at which they meet is the time required. This visualization of both events occurring on the same day is an example of **conceptual blending**.

Analogies and **metaphors** are types of conceptual blends. See also **identify**.

**Remark 1**  A conceptual blend is like an amalgamated sum or a pushout.

*References:* Conceptual blending, analogical mappings, metaphors and metonymies (these words overlap and different authors do not agree on their definitions) are hot topics in current cognitive science. These ideas have only just begun to be applied to the study of mathematical learning. See [Lakoff and Núñez, 1997], [Presmeg, 1997b]. More general references are [Fauconnier, 1997], [Fauconnier and Turner, 2002], [Katz *et al.*, 1998], [Lakoff, 1986], [Lakoff and Núñez, 1997], [Lakoff and Núñez, 2000].

*Acknowledgments:* The monk example is adapted from [Fauconnier, 1997], page 151.

**condition**    A **condition** is a requirement that occurs in the **definition** of a type of **mathematical object** or in the statement of a theorem or problem. See also **on**.

The word **requirement** is also used with this meaning. It means essentially the same thing as **constraint**, but the latter word seems to me occur mostly when the condition is an equation or an **inequality**. *Citations:* (205), (224), (316), (396), (428).

**conditional assertion**    A **conditional assertion** $A \Rightarrow B$ (pronounced $A$ **implies** $B$) is an **assertion** formed from two assertions $A$ and $B$, satisfying the following truth table:

| $A$ | $B$ | $A \Rightarrow B$ |
|:---:|:---:|:---:|
| T | T | T |
| T | F | F |
| F | T | T |
| F | F | T |

The operation "$\Rightarrow$" is called **implication**. It is sometimes written "$\supset$". In a given conditional $A \Rightarrow B$, $A$ is the **hypothesis** or the **antecedent** and $B$ is the **conclusion** or **consequent**. Warning: a conditional assertion is often called an implication, as well. *Citation:* (341).

In the **mathematical register**, $A \Rightarrow B$ may be written in several ways. Here are some examples where $A$ is "4 divides $n$" and $B$ is "2 divides $n$".

a) If 4 divides $n$, then 2 divides $n$. *Citations:* (79), (195), (410).
b) 2 divides $n$ if 4 divides $n$. *Citation:* (69)
c) 4 divides $n$ only if 2 divides $n$. *Citations:* (203), (387).
d) 4 divides $n$ implies 2 divides $n$. *Citations:* (156), (423).

e) Suppose [or Assume] 4 divides $n$. Then 2 divides $n$. *Citations:* (96), (163).

f) Let 4 divide $n$. Then 2 divides $n$. *Citation:* (353)

g) A **necessary** condition for 4 to divide $n$ is that 2 divide $n$. *Citations:* (195), (73).

h) A **sufficient** condition for 2 to divide $n$ is that 4 divide $n$. *Citations:* (195), (73).

i) The fact that $4 \mid n$ implies that $2 \mid n$. *Citation:* (423).

**Remark 1**   The word "if" in sentences (a), (b), and (c) can be replaced by "when" or (except for (c)) by "whenever". (See also **time**.)

**Remark 2**   Note that **if** has other uses, discussed under that word. The situation with **let**, **assume**, and **suppose** are discussed further in those entries.

**Remark 3**   Many other English constructions may be *translated* into (are equivalent to) conditional assertions. For example the **statement**
$$P \Leftrightarrow \text{"Every cyclic group is commutative"}$$
is equivalent to the statement "If $G$ is cyclic then it is commutative" (in a **context** where $G$ is of type "group"). But the statement $P$ is not itself a conditional assertion. See **universal quantifier**.

**Difficulties**   Students may have difficulties with implication, mostly because of **semantic contamination** with the usual way "if ... then" and "implies" are used in ordinary English. Some aspects of this are described here.

   In the first place, one way conditional sentences are used in ordinary English is to give rules. The effect is that "If $P$ then $Q$" means "$P$ if and only if $Q$".

**Example 1**   The sentence
   "If you eat your dinner, you may have dessert."

means in ordinary discourse that if you don't eat your dinner you may not have dessert. A child told this presumably interprets the statement as being in some sort of command mode, with different rules about "if" than in other types of sentences (compare the differences in the use of "if" in definitions and in theorems in the mathematical register.)

Perhaps as a consequence of the way they are used in ordinary English, students often take conditional sentences to be equivalences or even simply read them backward. Related fallacies are denying the hypothesis and affirming the consequent.

**Example 2**  A student may remember the fact "If a function is differentiable then it is continuous" as saying that being differentiable and being continuous are the same thing, or simply may remember it backward.

**Example 3**  When asked to prove $P \Rightarrow Q$, some students assume $Q$ and deduce $P$. This may have to do with the way students are taught to solve equations in high school.

Recommendation: if you state a mathematical fact in the form of a conditional assertion, you should *always* follow it by a statement explaining whether its converse is true, false or unknown. Besides providing useful additional information, doing this will remind the student about the direction of the conditional assertion.

**Remark 4**  Students have particular difficulty with only if and vacuous implication, discussed under those headings. See also false symmetry.

*References:* See [Fulda, 1989] for a discussion of many of the points in this entry. The sentence about dessert is from [Epp, 1995]. An analysis of conditionals in ordinary English is given by McCawley [1993], section 3.4 and Chapter 15. Other more technical approaches are in Section 2.1 of [Kamp and Reyle, 1993] and in Chapter 6 of [Chierchia and McConnell-Ginet, 1990].

**conjunction**  A **conjunction** is an **assertion** $P$ formed from two assertions $A$ and $B$ with the property that $P$ is **true** if and only if $A$ and $B$ are true. It is defined by the truth table shown here.

Do not confuse the usage of "conjunction" given here with the part of speech called "conjunction". Here, a conjunction is a whole **sentence**.

In the **mathematical register**, the conjunction of two assertions is usually signaled by connecting the two assertions with **and**. Examples are given under **and**.

| $A$ | $B$ | $P$ |
|-----|-----|-----|
| T | T | T |
| T | F | F |
| F | T | F |
| F | F | F |

**connective**  In mathematical logic, a **connective** or **logical connective** is a **binary operation** that takes pairs of **assertions** to an assertion. The connectives discussed in this text are **and**, **equivalent**, **imply**, and **or**. Note that some of these connectives are represented in English by conjunctions and others in more complex ways.

*Remark 1*  Unary operations such as **not** are sometimes called connectives as well. *Citations:* (148), (181).

**consider**  The command "Consider ... " introduces a (possibly variable) **mathematical object** together with **notation** for the objects and perhaps some of its structure.

*Example 1*  "Consider the function $f(x) = x^2 + 2x$." In the **scope** of this **statement**, $f(x)$ will denote that function specifically. *Citations:* (293), (302).

*One man's constant is another man's variable.*
   *–Alan Perlis*

**constant**  A specific **mathematical object** may be referred to as a **constant**, particularly if it is a number.

*Example 1*  The constant $\pi$ is the ratio of the circumference to the diameter of a circle, for any circle. *Citation:* (107).

More commonly, the word is used to refer to an object that may not be determined uniquely but that makes a statement containing various variables and parameters true for all values of the variables, but which may depend on the parameters.

**Example 2**

"There is a constant $K$ for which for any $x > K$, $x^2 > 100$."
The only variable is $x$ and there are no parameters. In this case $K$ is independent of $x$ but is not uniquely determined by the statement. *Citation:* (24).

**Example 3**

"There is a constant $K$ for which for any $x > K$, $x^2 + ax + b > 100$."
Here the statement contains two parameters and $K$ depends on both of them. Such a constant might better be called a "dependent parameter". *Citations:* (162), (187), (384).

A **constant function** is one for which $f(x) = f(y)$ for all $x, y \in$ dom $f$. *Citation:* (133).

See variable, parameter, unknown.

**constitute**   See collective plural.

**constraint**   A **constraint** is a requirement that occurs in the definition of a **mathematical object** or in the statement of a problem. Usually a constraint is an **equation** or an **inequality**, but almost any imposed requirement may be called a constraint. The words **condition** and **requirement** may also be used with a similar meaning.

**Example 1**   "Let $r$ be the root of a quadratic equation $x^2 + ax + b = 0$ subject to the constraint that $ab > 0$, that is, $a$ and $b$ have the same sign."

PI IS NOT JUST CONSTANT, BUT ALSO ETERNAL.

***Example 2***   "Let $p(x)$ be a polynomial subject to the constraint that all its roots are in the interval $[0, 1]$." *Citations:* (334), (363), (421)

**constructivism**   In mathematical education, this is the name given to the point of view that a student **constructs** her understanding of mathematical **concepts** from her experience, her struggles with the ideas, and what instructors and fellow students say. It is in opposition to the idea that the instructor in some sense pours knowledge into the student.

Of course, all I have given here are **metaphors**. However, constructivists draw conclusions concerning teaching and learning from their point of view.

"Constructivism" as a philosophy of education may connote other attitudes besides those discussed in the entry, including the idea that scientific knowledge does not or should not have a privileged position in teaching or perhaps in philosophy. My view in contrast to this is that in particular the **standard interpretation** of **mathematical discourse** should indeed have a privileged position in the classroom. This in no way implies that a student's private interpretation should be ridiculed. This is discussed further under **standard interpretation**.

***Remark 1***   Constructivism is the name of a point of view in the philosophy of mathematics as well, but there is no connection between the two ideas.

*References:* A brief description of constructivism in mathematics education may be found in [Selden and Selden, 1997]. Two very different expositions of constructivism are given by Ernest [1998] and Hersh [1997b]; these two books are reviewed in [Gold, 1999].

**contain**   If $A$ and $B$ are sets, the **assertion** $A$ contains $B$ can mean one of two things $B \subseteq A$ or $B \in A$. A similar remark can be made about the sentence "$B$ is contained in $A$."

***Remark 1***   Halmos, in [Steenrod *et al.*, 1975], page 40, recommends using "contain" for the membership relation and "**include**" for the inclusion relation. However, it appears to me that "contain" is used far more often to mean "include". *Citations:* (108), (142), (175), (181), (398).

The definition of "context" here gives a narrow meaning of the word and is analogous to its use in programming language semantics. The word has a broader meaning in ordinary **discourse**, typically referring to the physical or social surroundings.

**context** The **context** of an assertion includes the **interpretation** currently holding of the **identifiers** as well as any **assumptions** that may be in effect. **Definitions** and new assumptions change the context on the fly, so to speak. An experienced reader of **mathematical discourse** will be aware of the meanings of the various identifiers and assumptions and their changes as she reads.

**Example 1**  Before a phrase such as "Let $n = 3$", $n$ may be known only as an integer **variable**, or it may not have been used at all. After the phrase, it means specifically 3.

**Example 2**  A change of context may be given an explicit **scope**. For example, the assertion "In this chapter, we assume every group is Abelian" changes the context by restricting the interpretation of every **identifier** of **type** group.

**Example 3**  An **indefinite description** also changes the context.

> "On the last test I used a polynomial whose derivative had four distinct zeroes."

In the **scope** of such a sentence, **definite descriptions** such as "that polynomial" must refer specifically to the polynomial mentioned in the sentence just quoted.

**Remark 1**  The effect of each **assertion** in **mathematical discourse** can thus be interpreted as a function from context to context. This is described for one particular formalism (but not specifically for mathematical discourse) in [Chierchia and McConnell-Ginet, 1990], which has further references. See also [de Bruijn, 1994], page 875 and [Muskens, van Benthem and Visser, 1997].

## context-sensitive

*(a) Context-sensitive interpretation* The interpretation of a symbolic expression is **context-sensitive** if it depends on the place of the expression in the discourse containing it. The **pronunciation** of the expression may also vary with its place in the sentence.

*Example 1* If means if and only if when it occurs in a definition. This is discussed under if and **writing dilemma**.

*Example 2* In discourse concerning a **group** $G$, one might say both

"$G$ is commutative."

and

"Every odd number is an element of $G$."

In the first sentence, the reference is to the **binary operation** implicitly or explicitly given in the definition of "group", not to the **underlying set** of $G$. In the second, the reference is to the underlying set of the structure. Alternatively, one could say that the phrase "element of" is context sensitive: "element of $S$" means one thing if $S$ is a set and another if $S$ is a structure. In any case, the interpretation is context-sensitive.

That the **interpretation** of an expression is "context-sensitive" does not mean it depends on the **context** in the narrow meaning of the word given under that heading. In this situation, the "context" is the *syntactic* context of the expression, for example whether "if" occurs in a definition or elsewhere in Example 1, and whether the symbol $G$ occurs in the phrase "element of" in Example 2.

*(b) Context-sensitive pronunciation* The pronunciation of symbolic expressions in mathematics, particularly **symbolic assertions**, may depend on how they are used in the sentence. The most common way this happens is in the case of **parenthetic assertions**, under which examples are given.

## continuous

The notion of **continuous function** poses problems for students, even in the case of a **function** $f : A \to \mathbb{R}$ where $A$ is a subset of $\mathbb{R}$ (the set of all **real** numbers). In that case, it is defined using the standard $\epsilon$-$\delta$ definition. This can be given an **expansive generalization** to metric **spaces**, and even

further a **reconstructive generalization** to general topological spaces via
the rule about inverse images of open sets (quite baffling to some students).

**(a) *Continuity is several related ideas*** One can define what it means
for a function

- to be continuous at a point,
- to be continuous on a set
- to be continuous (that is, on its whole domain).

The relation among these ideas has subtleties; for example the function
$1/x$ is continuous at every point at which it is defined, but not continuous
on $\mathbb{R}$.

**(b) *Metaphors for continuity are usually inaccurate*** The various
**intuitions** for continuity that students hear about are mostly incorrect.

***Example 1*** The well-known unbroken-graph intuition fails for functions such as

$$f(x) = \begin{cases} \sin\left(\frac{1}{x}\right) & x \neq 0 \\ 0 & x = 0 \end{cases}$$

This function is not continuous at 0 but nowhere is there a "break" in
the graph.

**(c) *Continuity is hard to put in words*** It is difficult to say precisely
in words what continuity means. The $\epsilon$-$\delta$ definition is logically complicated with nested **quantifiers** and several **variables**. This makes it difficult
to understand. and attempts to put the $\epsilon$-$\delta$ definition into words usually
fail to be accurate. For example:

> "A function is continuous if you can make the output change
> as little as you want by making the change in the input small
> enough."

That paraphrase does not capture the subtlety that if a certain change in the input will work, then so must any smaller change at the same place. See also **map**.

> *Citations:* (164), (200), (230), (233),
> *References:* This discussion draws from [Exner, 2000], Chapter 2. See also [Núñez and Lakoff, 1998] and [Núñez, Edwards and Matos, 1999].

**continuum hypothesis**    In mathematics, the **continuum hypothesis** is the statement that there is no cardinality between that of the integers and that of the reals. In fluid dynamics and elsewhere, it is used to mean that discrete data can be usefully modeled by a continuous function. These two meanings are independent of each other. I suspect that the second meaning is the product of **semantic contamination** by the first meaning, which dates as early as 1927. *Citations:* (35), (117), (182), (412).

**contrapositive**    The **contrapositive** of a **conditional assertion** $P \Rightarrow Q$ is the statement (not $Q$) $\Rightarrow$ (not $P$). In mathematical **arguments**, the conditional assertion and its contrapositive are equivalent. In particular, to prove $P \Rightarrow Q$ it is enough to prove that (not $Q$) $\Rightarrow$ (not $P$), and once you have done that, *no further argument is needed*. I have attended lectures where further argument *was* given, leading me to suspect that the lecturer did not fully understand the contrapositive, but I have not discovered an instance in print that would indicate that. See **proof by contradiction**.

**Remark 1**    The fact that a conditional assertion and its contrapositive are logically equivalent means that a proof can be organized as follows, and in fact many proofs in texts are organized like this:

  a) Theorem: $P$ implies $Q$.

  b) Assume that $Q$ is false.

c) Argument that not $P$ follows.

d) Conclude that $P$ implies $Q$.

e) End of proof.

Often $P$ is a conjunction of several statements $P_1, \dots P_n$ and the argument in the third step will be an argument that not $P_i$ for some particular $i$.

The reader may be given no hint as to the form of the proof; she must simply recognize the pattern. A concrete example of such a proof is given under functional knowledge. See also pattern recognition.

***Difficulties*** In contrast to the situation in mathematical reasoning, the contrapositive of a conditional sentence in ordinary English about everyday topics of conversation does not in general mean the same thing as the direct sentence. This causes semantic contamination.

***Example 1*** The sentence

"If it rains, I will carry my umbrella."

does not mean the same thing as

"If I don't carry my umbrella, it won't rain."

There are reasons for the difference, of course, but teachers rarely explain this to students. McCawley [1993], section 3.4 and Chapter 15, discusses the contrapositive and other aspects of conditional sentences in English. More about this in the remarks under only if.

*Citations:* (208), (413).

**convention**    A convention in mathematical discourse is notation or terminology used with a special meaning in certain contexts or in certain fields.

***Example 1*** The use of if to mean "if and only if" in a definition is a convention. This is controversial and is discussed under if.

***Example 2*** Constants or parameters are conventionally denoted by $a$, $b$, ..., functions by $f$, $g$, ... and variables by $x$, $y$, ....

***Example 3*** Referring to a group (or other mathematical structure) and its underlying set by the same name is a convention.

***Example 4*** The meaning of $\sin^n x$ is the inverse sine (arcsin) if $n = -1$ but the multiplicative power for positive $n$ ($\sin^n x = (\sin x)^n$). This is a common convention in calculus texts, usually explicit.

***Remark 1*** Example 1 exhibits context-sensitivity. Examples 3 and 4 exhibit failure of compositionality. Example 4 is not an example of context-sensitivity since the meaning depends on what $n$ itself is.

***Remark 2*** Examples 1 through 4 differ in how pervasive they are and in whether they are made explicit or not. The convention in Example 1 is so pervasive it is almost never mentioned (it is just beginning to be mentioned in textbooks aimed at first courses in abstract mathematics). That is almost, but not quite, as true of the second convention. The third and fourth conventions are quite common but often made explicit.

Any given culture has some customs and taboos that almost no one inside the culture is aware of, others that only some who are particularly sensitive to such issues (or who have traveled a lot) are aware of, and still others that everyone is aware of because it is regarded as a mark of their subculture (such as grits in the American south). One aspect of this Handbook is an attempt to uncover features of the way mathematicians talk that mathematicians are not generally aware of.

***Example 5*** Some conventions are pervasive among mathematicians but different conventions hold in other subjects that use mathematics. An example is the use of $i$ to denote the imaginary unit. In electrical engineering it is commonly denoted $j$ instead, a fact that many mathematicians are unaware of. I first learned about it when a student asked me if $i$ was the same as $j$. *Citation:* (332).

***Example 6*** Other conventions are pervasive in one country but may be different in another. See part (b) of trigonometric functions for examples.

See also positive, radial concept and the sidebar under real number.

**converse**   The **converse** of a conditional assertion $P \Rightarrow Q$ is $Q \Rightarrow P$. Students often fall into the trap of assuming that if the assertion is **true** then so is its converse; this is the fallacy of **affirming the consequent**. See also **false symmetry**.

**coordinatewise**   A function $F : A \to B$ induces a function often called $F^*$ from lists of elements of $A$ to lists of elements of $B$ or from $A^n$ to $B^n$ for a fixed **positive integer** $n$ by defining

$$F^*(a_1, \dots, a_n) = (F(a_1), \dots, F(a_n))$$

One says that this defines $F^*$ **coordinatewise** or **componentwise**.

**Example 1**   "In the product of two groups, multiplication is defined coordinatewise."

    One can say that assertions are defined coordinatewise, as well. (See Remark 1 under **assertion**.)

**Example 2**   "The product of two ordered sets becomes an ordered set by defining the order relation coordinatewise."

    *Citations:* (68), (244), (54).

**copy**   When one is discussing a **mathematical structure**, say the ring of **integers**, one sometimes refers to "a copy of the integers", meaning a structure isomorphic to the integers. This carries the connotation that there is a preferred copy of the **mathematical object** called the integers (see **specific mathematical object**); I suspect that some who use this terminology don't believe in such preferred copies. Our language, with its **definite descriptions** and proper nouns, is not particularly suited to discussing things defined unique **up to isomorphism**. *Citations:* (398) (111), (144).

**coreference**   **Coreference** is the use of a word or phrase in **discourse** to denote the same thing as some other word or phrase. In English, third person pronouns (he, she, it, they), demonstratives (this, that, these, those), and the word "do" are commonly used for coreference.

In this entry I will discuss two aspects of coreference that has caused confusion among my students.

**(a) *Collective coreference*** Some years ago the following question appeared in my classnotes [Wells, 1997]:

> "Cornwall Computernut has 5 computers with hard disk drives and one without. Of these, several have speech synthesizers, including the one without hard disk. Several have Pascal, including those with synthesizers. Exactly 3 of the computers with hard disk have Pascal. How many have Pascal?"

Linguists have formulated some of the rules that govern the use of coreference in English. Typically, the rules produce some syntactic restrictions on what can be referred to, which in some cases determine the reference uniquely, but in other cases the meaning must be left ambiguous to be disambiguated (if possible) by the situation in which it is uttered.

The phenomenon of coreference is also called **anaphora**, a word borrowed from rhetoric which originally meant something else. Many (but not all) linguists restrict "anaphora" to backward coreference and use "cataphora" for forward reference. Some linguists call forward reference "backward dependency". I took the name "forward reference" from computing science.

Some students did not understand that the phrase "including those with synthesizers" meant "including *all* those with synthesizers" (this misunderstanding removes the uniqueness of the answer). They were a minority, but some of them were quite clear that "including those with synthesizers" means *some or all* of those with synthesizers have Pascal; if I wanted to require that all of them have Pascal I would have to say "including all those with synthesizers". A survey of a later class elicited a similar minority response. This may be related to common usage in **setbuilder notation**. *Citations:* (206), (225), (226).

I do not know of any literature in linguistics that addresses this specific point.

**(b) *Forward reference*** A **forward reference** occurs when a pronoun refers to something named later in the text.

***Example 1***   This is a problem I gave on a test:

"Describe how to tell from its last digit in base 8 whether an integer is even."

In this sentence "its" refers to "an integer", which occurs later in the sentence.

***Remark 1***   That problem and other similar problems have repeatedly caused a few of my students to ask what it meant. These included native English speakers. Of course, this problem is not specific to mathematical discourse.

*References:* Introductions to the topic are in [Harley, 2001], pages 322–325 and [Fiengo and May, 1996]. See also [Kamp and Reyle, 1993], pp. 66ff, [Chierchia, 1995], [McCarthy, 1994], [Halliday, 1994], pp. 312ff.

See also respectively.

**corollary**   A corollary of a theorem is a fact that follows easily from the theorem. *Citations:* (56), (79), (217), (313) (corollary of two theorems).

***Remark 1***   "Easily" may mean by straightforward calculations, as in Citation (56), where some of the necessary calculations occur in the proof of the theorem, and in Citation (313), or the corollary may be simply an instance of the theorem as in Citation (217).

**counterexample**   A counterexample to an universally quantified assertion is an instance of the assertion for which it is false.

***Example 1***   A counterexample to the assertion

"For all real $x$, $x^2 > x$"

is $x = 1/2$. See also example.

*Citations:* (59), (205).

**counting number**   The **counting numbers** may denote the positive integers, the nonnegative integers, or apparently even all the integers (although I don't have an unequivocal citation for that).

I have also heard people use the phrase to denote the number of mathematical objects of a certain type parametrized by the positive or nonnegative integers. For example, the $n$th Catalan number can be described as the counting number for binary trees with $n + 1$ leaves. *Citation:* (350).

**covert curriculum**   The **covert curriculum** (or **hidden curriculum**) consists of the skills we expect math students to acquire without our teaching the skills or even mentioning them. What is in the covert curriculum depends to some extent on the teacher, but for students in higher level math courses it generally includes the ability to read mathematical texts and follow mathematical proofs. (We do try to give the students explicit instruction, usually somewhat offhandedly, in how to *come up with* a proof, but generally not in how to read and follow one.) This particular skill is one that this Handbook is trying to make overt. There are undoubtedly other things in the covert curriculum as well.

*Reference:* [Vallance, 1977].

*Acknowledgments:* I learned about this from Annie Selden. Christine Browning provided references.

**crisp**   See concept.

**curly brackets**   See brace.

**dash**   See prime.

**decreasing**   See increasing.

**default**   An interface to a computer program will have various possible choices for the user to make. In most cases, the interface will use certain choices automatically when the user doesn't specify them. One says the program **defaults** to those choices.

***Example 1***   A word processing program may default to justified paragraphs and insert mode, but allow you to pick ragged right and typeover mode.

The concept of default is a remarkably useful one in linguistic contexts. For example, there is a sense in which the word "ski" defaults to snow skiing in Minnesota and to water skiing in Georgia. Similarly "CSU" defaults to Cleveland State University in northern Ohio and to Colorado State University in parts of the west.

Default usage may be observed in many situations in **mathematical discourse**. Some examples from my own experience:

> I have spent a lot of time in both Minnesota and Georgia and the remarks about skiing are based on my own observation. One wonders where the boundary line is. Perhaps people in Kentucky are confused on the issue!
>
> These usages are not absolute. Some affluent Georgians (including *native* Georgians) may refer to snow skiing as "skiing", for example, and this usage can be intended as a kind of snobbery.

***Example 2***   To algebraists, the "free group" on a set $S$ is non-Abelian. To some topologists, this phrase means the free *Abelian* group.

***Example 3***   In informal conversation among some analysts, functions are automatically continuous.

***Example 4***   "The group $\mathbb{Z}$" usually means the group with $\mathbb{Z}$ (the set of integers) as **underlying set** and addition as operation. There are of course many other group operations on $\mathbb{Z}$. Indeed, the privileged nature of the addition operation may be part of a mathematician's **schema** for $\mathbb{Z}$.

> This meaning of "default" has made it into dictionaries only in the last ten years. This usage does not carry a derogatory connotation.

See also **theory of functions** and **number theory**.

**defining condition**   See **setbuilder notation**.

**defining equation**    See function.

**definite article**    The word "the" is called the **definite article**. It is used in forming definite descriptions.

*(a) The definite article as universal quantifier*    Both the indefinite article and the definite article can have the force of universal quantification. Examples are given under universal quantifier.

*(b) The definite article and setbuilder notation*    A set $\{x \mid P(x)\}$ in setbuilder notation is often described with a phrase such as "the set of $x$ such that $P(x)$". In particular, this set is the set of *all* $x$ for which $P(x)$ is true.

*Example 1*    The set described by the phrase "the set of even integers" is the set of *all* even integers.

*Difficulties*    Consider this test question:
> "Let $E$ be the set of even integers. Show that the sum of any two elements of $E$ is even."

Students have given answers such as this:
> "Let $E = \{2, 4, 6\}$. Then $2 + 4 = 6$, $2 + 6 = 8$ and $4 + 6 = 10$, and 6, 8 and 10 are all even."

This misinterpretation has been made in my classes by both native and non-native speakers of English.

*(c) Definite article in definitions*    The definiendum of a definition may be a definite description.

*Example 2*    "The sum of vectors $(a_1, a_2)$ and $(b_1, b_2)$ is $(a_1 + b_1, a_2 + b_2)$." I have known this to cause difficulty with students in the case that the definition is not clearly marked as such. The definite description makes the student believe that they should know what it refers to. In the assertion in this example, the only clue that it is a definition is that "sum" is in **boldface**. This is discussed further under **definition**. *Citations:* (40), (101), (150).

**definite description**   A noun phrase in which the **determiner** is "the" or certain other words such as "this", "that", "both", and so on, is called an **definite description** or a **definite** noun phrase. Such a phrase refers to a presumably uniquely determined **object**. The assumption is that the object referred to is already known to the speaker and the listener or has already been referred to.

**Example 1**  If you overheard a person at the blackboard say to someone
 "The function is differentiable, so ... "
you would probably assume that that person is referring to a function that speaker and listener both already know about. It may be a specific **function**, but it does not have to be; they could be in the middle of a proof of a theorem about functions of a certain type and "the function" could be a **variable** function that they named for the purposes of proving the theorem.

    This example shows that in the **mathematical register**, whether a description is definite or indefinite is independent of whether the **identifier** involved is **determinate** or **variate**.

**Example 2**  "Let $G$ be a **group**. Show that the identity of $G$ is idempotent." This example shows that the presumptive uniquely determined **object** ("the identity") can depend on a **parameter**, in this case $G$.

**Example 3**  The phrase "the equation of a plane" is a definite description with a parameter (the plane).
    *Citations:* (170), (133).

    See [Kamp and Reyle, 1993], Section 3.7.5.

## definition

### 1. Mathematical definitions

A **mathematical definition** prescribes the meaning of a **symbol**, word, or phrase, called the **definiendum** here, as a **mathematical object** satisfying all of a set of requirements. The definiendum will be either an adjective that denotes a property that **mathematical objects** may have, or it may be a **noun phrase** that denotes a type of mathematical object with certain properties.

> A mathematical definition is fundamentally different from other sorts of definitions, a fact that is not widely appreciated by mathematicians. The differences are dicussed under **concept** and under **dictionary definition**.

Mathematical texts sometimes define other parts of speech, for example in the case of **vanish**, but that possibility will not be discussed here.

**(a) Syntax of mathematical definitions** Definitions of nouns and of adjectives have similar syntax, with some variations. Every definition will contain a **definiendum** and a **definiens**, which is a set of properties an object must have to be correctly named by the definiendum. The definiens may be syntactically scattered throughout the definition much as the Union Terriroty of Pondicherry is scattered throughout India.

In particular, a definition may have any or all of the following structures:

1. A **precondition**, occurring before the definiendum, which typically gives the type of structure that the definition applies to and may give other conditions.

2. A **defining phrase**, a list of conditions on the definiendum occurring in the same **sentence** as the definiendum.

3. A **postcondition**, required conditions occurring after the ostensible definition which appear to be an afterthought. The postcondition commonly begins with "**where**" and some examples are given under that heading.

66

*(i) Direct definitions*   One can define "domain" in point set topology directly by saying

"A **domain** is a connected open set."

(See **be**.) The **definiendum** is "domain" and the defining phrase (which constitutes the entire **definiens**) is "is a connected open set". Similarly:

"'An **even integer** is an integer that is divisible by 2."

*Citations:* (40), (55). In both these cases the definiendum is the subject of the sentence.

**Remark 1**   The definition of "domain" given in the preceding paragraph involves a **suppressed parameter**, namely the **ambient** topological space.

*(ii) Definitions using conditionals*   It is more common to word the definition using "if", in a **conditional** sentence. In this case the subject of the sentence is a **noun phrase** giving the **type** of **object** being defined and the **definiendum** is given in the predicate of the **conclusion** of the conditional sentence. The subject of the sentence may be a **definite noun phrase** or an **indefinite** one. The conditional sentence, like any such, may be worded with hypothesis first or with conclusion first. All this is illustrated in the list of examples following, which is not exhaustive.

1. [Indefinite noun phrase, **definiendum** with no proper name.] A set is a **domain** if it is open and connected. Or: If a set is open and connected, it is a **domain**. Similarly: An integer is **even** if it is divisible by 2. *Citation:* (361).

2. [Indefinite noun phrase, **definiendum** given proper name.] A set $D$ is a **domain** if $D$ is open and connected. An integer $n$ is **even** if $n$ is divisible by 2. (In both cases and in similar named cases below the second occurrence of the name could be replaced by "it".) *Citations:* (156), (222).

3. [Definite noun phrase.] The set $D$ is a **domain** if $D$ is open and connected. Similarly: The integer $n$ is **even** if $n$ is divisible by 2. Using the definite form is much less common than using the indefinite form, and seems to occur most often in the older literature. It requires that the definiendum be given a proper name. *Citations:* (14), (21).

4. [Using "let" in a precondition.] Let $D$ be a set. Then $D$ is a **domain** if it is open and connected. Similarly: Let $n$ be an integer. Then $n$ is **even** if it is divisible by 2. *Citation:* (124).

5. [Using "if" in a precondition] If $n$ is an integer, then it is **even** if it is divisible by 2. *Citation:* (40).

**Remark 2** All the definitions above are given with the definiendum marked by being in **boldface**. Many other forms of marking are possible; see marking below.

A symbolic expression may be defined by using phrases similar to those just given.

**Example 1** "For an integer $n$, $\sigma(n)$ is the sum of the positive divisors of $n$."

Sometimes "define" is used instead of "let" in the sense of "assume".

**Example 2** "Define $f(x)$ to be $x^2 + 1$. What is the derivative of $f$?"

Students sometimes wonder what they are supposed to do when they read a sentence such as "Define $f(x)$ to be $x^2 + 1$", since they take it as a command. *Citations:* (237), (242).

Other ways of giving a definition use call, put, say and set, usually in the imperative the way "define" is used in Example 2. Many other forms of syntax are used, but most of them are either a direct definition or a definition using a conditional, with variations in syntax that are typical of academic prose.

***Remark 3***  Some authors have begun using "if and only if" in definitions instead of "if". More about this in the entry for if and in writing dilemma. See also convention and let.

   ***(iii) Marking the definiendum***  The definiendum may be put in italics or quotes or some other typeface instead of **boldface**, or may not be **marked** at all. When it is not marked, one often uses signaling phrases such as "is defined to be", "is said to be", or "is called", to indicate what the definiendum is. A definition may be delineated, with a label "Definition". *Citations:* (156), (211) (formally marked as definition); (261), where it is signaled as definition by the sentence beginning "We call two sets ..."; (126) and (139), where the only clue that it is a definition is that the word is in boldface; (55) and (218), where the clue is that the word is in italics.

***Remark 4***  Words and phrases such as "We have defined..." or "recall" may serve as a valuable clue that what follows is *not* a definition.

***Remark 5***  Some object to the use of boldface to mark the definiendum. I know of no such objection in print; this observation is based on my experience with referees.

***(b) Mathematical definitions and concepts***  The definition of a concept has a special logical status. It is the fundamental fact about the concept from which all other facts about it must ultimately be deduced. I have found this special logical status one of the most difficult concepts to get across to students beginning to study abstract mathematics (in a first course in linear algebra, discrete mathematics or abstract algebra). There is more about this under concept, mental representation, rewrite using definitions and trivial. See also unwind.

   There is of course a connection among the following three ideas:
a) The uses of the word function in the mathematical register.
b) The mathematical definition of function.

c) The mental representation associated with "function".

To explicate this connection (for all mathematical concepts, not just "function") is a central problem in the philosophy of mathematics.

*References:* [Lakoff and Núñez, 2000], Chapter 5, [Bills and Tall, 1998], [Tall and Vinner, 1981], [Vinner, 1992], [Vinner and Dreyfus, 1989], [Wood, 1999].

*(c) Definition as presentation of a structure* A mathematical definition of a concept is spare by intent: it will generally provide an irredundant list of data and of relationships that must hold among the data. Data or properties that follow from other given items are generally not included intentionally in a definition (some exceptions are noted under redundant) and when they are the author may feel obligated to point out the redundancy.

As a result, a mathematical definition hides the richness and complexity of the concept and as such may not be of much use to students who want to understand it (gain a rich mental representation of it). Moreover, a person not used to the minimal nature of a mathematical definition may gain an exaggerated idea of the importance of the items that the definition *does* include. See also literalist.

### 2. Dictionary definition

An explanation, typically in a dictionary or glossary, of the meaning of a word. This is not the same as a mathematical definition (meaning (1) above). To distinguish, this Handbook will refer to a definition of the sort discussed here as a **dictionary definition**. The entries in this Handbook are for the most part dictionary definitions.

*Example 3* The entry for "function" given in this Handbook describes how the word "function" and related words are used in the **mathematical register**. The definition of function given in a typical mathematical textbook (perhaps as a set of ordered pairs with certain properties) spec-

*DICTIONARY, n. A malevolent literary device for cramping the growth of a language and making it hard and inelastic. THIS dictionary, however, is a most useful work.*
   *–Ambrose Bierce, The Devil's Dictionary.*

ifies what kind of mathematical object is to be called a function. See Remark 2 under free variable for a discussion of this issue in a particular case.

*Acknowledgments:* Atish Bagchi

## definition by cases    See cases.

## degenerate    An example of a type of mathematical structure is in some disciplines called **degenerate** if either

  (i)  some parts of the structure that are distinct in the definition of that type coincide (I call this **collapsing**), or

 (ii)  some parameter is zero.

The converse, that if a structure satisfies (i) or (ii) then it is called degenerate, is far from being correct; the word seems to be limited to certain specific disciplines.

*Example 1*   A line segment can be seen as a degenerate isosceles triangle – two sides coincide and the third has zero length. Note that this fits both (i) and (ii).

*Example 2*   The concept of degenerate critical point has a technical definition (a certain matrix has zero determinant) and is responsible for a sizeable fraction of the occurrences of "degenerate" I found on JSTOR. A small

> The definition of degenerate given here is based on reading about thirty examples of the use of the word on JSTOR. Sometimes the word has a mathematical definition specific to the particular discipline of the paper and sometimes it appears to be used informally.

perturbation turns a degenerate critical point into several critical points, so this can be thought of as a kind of collapsing.

*Citations:* (343), (359).
*Acknowledgments:* Robin Chapman.

## degree    See order and trigonometric functions.

## delimiter    **Delimiters** consist of pairs of symbols used in the symbolic language for grouping.

*(a) Bare delimiters* A pair of delimiters may or may not have significance beyond grouping; if they do not they are **bare delimiters**. The three types of **characters** used as bare delimiters in mathematics are parentheses, square brackets, and braces.

Typically, parentheses are the standard delimiters in symbolic expressions. Square brackets or braces may be used to aid parsing when parentheses are nested or when the expression to be enclosed is large, but square brackets and braces are occasionally used alone as bare delimiters as well.

**Example 1** The **expression** $\left((x+1)^2 - (x-2)^2\right)^n$ contains nested parentheses and might alternatively be written as $\left[(x+1)^2 - (x-2)^2\right]^n$.

Parentheses, square brackets and braces may also be used with additional significance; such uses are discussed with examples under their own headings.

I have been unable to find a citation for the use of angle brackets as *bare* delimiters, although of course they are commonly used as delimiters that carry meaning beyond grouping.

*(b) Other delimiters* Other symbols also are used to carry meaning and also act as delimiters. Examples include:

- The symbol for absolute value, as in $|x|$.
- The symbol for the norm, as in $\|x\|$.
- The integral sign, discussed under **integral**.
  *Citations:* (103), (275), (380), (381).

**delineated** A piece of **text** is **delineated** if it is set off typographically, perhaps as a **display** or by being enclosed in a rectangle. Delineated text is often **labeled**, as well.

*Example 1* "**Theorem** *An integer n that is divisible by 4 is divisible by 2.*" The label is "Theorem".

**denote**  To say that an expression $A$ **denotes** a specific **object** $B$ means that $A$ refers to $B$; in an **assertion** containing a description of $B$, the description can be replaced by $A$ and the truth value of the assertion remains the same. $B$ is sometimes called the **denotation** of $A$.

*Example 1*  The symbol $\pi$ denotes the ratio of the circumference of a circle to its diameter.

> *Citations:* (342), (101).

*Remark 1*  [Krantz, 1997], page 38, objects to the use of "denote" when the expression being introduced refers (in my terminology) to a **variable mathematical object**, for example in a sentence such as "Let $f$ denote a continuous function".

*Remark 2*  Some authors also object to the usage exemplified by "the ratio of the circumference of a circle to its diameter is denoted $\pi$"; they say it should be "denoted by $\pi$". *Citations:* (176), (251). (333), (388), (124).

**denying the hypothesis**  The fallacy of deducing not $Q$ from $P \Rightarrow Q$ and not $P$. Also called **inverse error**.

*Example 1*  You are asked about a certain subgroup $H$ of a non-abelian group $G$. You "know" $H$ is not normal in $G$ because you know the theorem that if a group is Abelian, then every subgroup is normal in it.

> In contrast, consider Example 1 under **conditional assertion**.

**dependency relation**  See function.

**dependent variable notation**  This is a method of referring to a function that uses the pattern

> "Let $y$ be a function of $x$."

where $x$ is an identifier for the input and $y$ is an identifier for the output. In this case, one also says that $y$ is **depends** or is **dependent** on $x$. The rule for the function may not be given.

In this usage, the value of the unnamed function at $x$ is sometimes denoted $y(x)$. Note that this does not qualify as **structural notation** since the notation does not determine the function. *Citations:* (239), (384).

**deprecate**   The word **deprecate** is used in this Handbook to refer to a usage occuring in **mathematical discourse** that one could reasonably say should not be used for some good reason (usually because it causes unnecessary confusion). This is sometimes my opinion and sometimes a reference to another author's opinion.

I have borrowed this word from computing science.

**determinate**   A free identifier is **determinate** if it refers to a specific **mathematical object**.

***Example 1***   The symbol "3" is determinate; it refers to the unique integer 3. But see Remark (a) under **mathematical object**.

An extended discussion of determinate and variate identifiers may be found under **variate**.

**discourse**   Connected meaningful speech or writing. Connected meaningful *writing* is also called **text**.

**Discourse analysis** is the name for the branch of linguistics that studies how one extracts meaning from sequences of sentences in natural language. [Kamp and Reyle, 1993] provides a mathematical model that may explain how people extract logic from connected discourse, but it does not mention the special nature of mathematical exposition. A shorter introduction to discourse analysis is [van Eijck and Kamp, 1997].

**disjoint**   Two sets are **disjoint** if their intersection is empty.

*Example 1*    "$\{1,2\}$ and $\{3,4,5\}$ are disjoint."

The word may be used with more than two sets, as well:

*Example 2*    "Let $\mathcal{F}$ be a family of disjoint sets."

*Example 3*    "Let $A$, $B$ and $C$ be disjoint sets."

       *Citations:* (143), (323).

**Difficulties** Students sometimes say things such as: "Each set in a partition is disjoint". This is an example of a missing relational argument (see Section (iii) under **behaviors**).

**disjunction**    A **disjunction** is an **assertion** $P$ formed from two **assertions** $A$ and $B$ with the property that $P$ is **true** if and only if at least one $A$ and $B$ is true. It is defined by the following truth table:

| $A$ | $B$ | $P$ |
|-----|-----|-----|
| T | T | T |
| T | F | T |
| F | T | T |
| F | F | F |

In the **mathematical register**, the disjunction of two assertions is usually signaled by connecting the two assertions with "**or**". Difficulties with disjunctions are discussed under **or**.

**disjunctive definition**    See **cases**.

**display**    A **symbolic expression** is **displayed** if it is put on a line by itself. Displays are usually centered. The word "displayed" is usually used only for symbolic expressions. See **delineated**.

**distinct**    When several new **identifiers** are introduced at once, the word **distinct** is used to require that no two of them can have the same value.

*Example 1*   "Let $m$ and $n$ be distinct integers."

This means that in the following **argument**, one can assume that $m \neq n$.

*Difficulties*   Students may not understand that without a word such as "distinct", the **variables** may indeed have the same value. Thus

"Let $m$ and $n$ be integers."

allows $m = n$. In [Rota, 1996], page 19, it is reported that E. H. Moore was sufficiently bothered by this phenomenon to say,

"Let $m$ be an integer and let $n$ be an integer."

*Citations:* (238), (196), (279).

**distributive plural**   The use of a plural as the subject of a **sentence** in such a way that the predicate applies individually to each item referred to in the subject.

*Example 1*   "The multiples of 4 are even." (or "All the multiples of 4 are even" – see **universal quantifier**.)

This phenomenon is given a theoretical treatment in [Kamp and Reyle, 1993], pages 320ff. See also **collective plural** and **each**. *Citations:* (19), (225), (226).

## divide

### 1. Divides for integers

An **integer** $m$ **divides** an integer $n$ (or: $m$ is a **divisor** or **factor** of $n$) if there is an integer $q$ for which $n = qm$. Some authors require that $q$ be uniquely determined, which has the effect of implying that no integer divides 0. (0 does not divide any other integer in any case.) This definition, with or without the requirement for uniqueness, appears to be standard in texts in discrete mathematics and **number theory**. *Citations:* (69), (79).

## 2. Divides for commutative rings

If $a$ and $b$ are elements of a commutative ring $R$, then $a$ divides $b$ if there is an element $c$ of $R$ with the property that $b = ac$. This appears to be the standard definition in texts in **abstract algebra**. I am not aware of any such text that requires uniqueness of $c$.

Of course, the second meaning is a **generalization** of the first one. I have known this to cause people to assert that every nonzero integer divides every integer, which of course is true in the second meaning, taking the commutative ring to be the ring of rationals or reals. *Citation:* (76).

**domain**   The **domain** of a function must be a **set** and may be named in any way that sets are named. The domain is frequently left unspecified. It may be possible to deduce it from what is stated; in particular, in cases where the **rule** of the function is a **symbolic expression** the domain may be implicitly or explicitly assumed to be the set of all values for which the expression is defined.

> Most authors require that a function be defined at every **element** of the domain, if the domain is specified. A **partial function** is a **mathematical object** defined in the same way as a function except that it may be defined for only a subset of the domain.

**Remark 1**   The set of values for which the expression is defined is a subtle idea. Consider: Let $x$ be a real variable. Is the expression $x \tan(x + \pi/2)$ defined at $x = 0$?

- If you say all parts have to be meaningful, it is not defined. This is what is called **eager evaluation** in computing science.
- If you start evaluating it from left to right and come up with a definite value before you have considered all parts of the expression, then the value is 0 (this is **lazy evaluation**).

*Citations:* (174), (384).

*Acknowledgments:* [Miller, 2003].

**(a) Notation for domain**   Aside from **straight arrow notation** the following phrases may be used to state that a set $S$ is the domain of a function $f$:

   a) $f$ is a function with domain $S$. *Citation:* (133).

   b) $\mathrm{dom}\, f = S$. *Citations:* (122), (170).

   c) $f$ is a function on $S$. *Citations:* (26), (42), (188).

See also **defined in**.

***Remark 2*** The word "domain" is also used in topology (connected open set) and in computing (lattice satisfying various conditions) with meanings unrelated to the concept of domain of a function. See **multiple meanings**. *Citation:* (205).

**dummy variable**   Same as bound variable.

**e.g.**   See i.e.

**each**   Generally can be used in the same way as **all**, **every**, and **any** to form a **universal quantifier**.

***Example 1***   "Each multiple of 4 is even."

***Remark 1***   It appears to me that this direct use of "each" is uncommon. When it is used this way, it always indicates a **distributive plural**, in contrast to **all**.

    "Each" is more commonly used before a noun that is the object of a preposition, especially after "for", to have the same effect as a distributive plural.

***Example 2***   "For each even number $n$ there is an integer $k$ for which $n = 2k$."

***Example 3***   "A *binary operation* $*$ *on a set* is a rule that assigns to each ordered pair of elements of the set some element of the set." (from [Fraleigh, 1982], page 11).

***Example 4*** Some students do not grasp the significance of a postposited "each" as in the sentence below.

"Five students have two pencils each."

This means that each of the five students has two pencils (a different two for each student). This usage occurs in combinatorics, for example.

*Citations:* (66), (110), (214), (369).

**element** If $S$ is a set, the expression "$x \in S$" is pronounced in English in any of the following ways:

a) "$x$ is in $S$". *Citations:* (126), (133).

b) "$x$ is an **element** of $S$" [or "in $S$"]. *Citations:* (196), (369).

c) "$x$ is a **member** of $S$". *Citation:* (41).

d) "$S$ contains $x$" or "$x$ is contained in $S$". *Citations:* (142), (108).

***Remark 1*** Sentence (d) could also mean $x$ is contained in $S$ as a subset. This is not likely to cause confusion because of the common practice of writing sets with **uppercase** letters and their elements as **lowercase**. See **contain**.

***Remark 2*** A common **myth** among students is that there are two kinds of **mathematical objects**: "sets" and "elements". This can cause confusion when they are faced with the idea of a set being an element of a set. The word "element" is used by experienced mathematicians only in a phrase involving both a **mathematical object** and a set. In particular, being an element is not a property that some **mathematical objects** have and some don't.

*Acknowledgments:* Atish Bagchi

**elementary** In everyday English, an explanation is "elementary" if it is easy and if it makes use of facts and principles known to most people. Mathematicians use the word "elementary" with other meanings as well. Most of them are technical meanings in a specific type of mathematics. We consider two uses in mathematicans' **informal jargon**.

*(a)* *Elementary proofs*  A proof of a theorem is **elementary** if it uses only ideas from the same field as the theorem. Rota [Rota, 1996], pages 113 ff., discusses the case of the prime number theorem in depth; the first proofs around 1900 used complex function theory, but it was given an elementary proof much later. That proof was quite long and complicated, not at all elementary in the non-mathematician's sense. (A simpler one was found much later.)

*(b)* *Elementary definitions*  Mathematicians sometimes use "elementary" in another sense whose meaning is not quite clear to me. It is apparently in opposition to **conceptual**. Here are two possible definitions; we need **citations** to clear this up.

a) A **definition** of a type of **mathematical structure** is elementary if it involves quantifying only over the elements of the **underlying set**(s) of the structure. In particular it does not involve quantifying over sets or over functions. This is the meaning used by Vought [1973], page 3.

b) A definition of a type of structure is elementary if it does not make use of other definitions at the same level of abstraction. Thus it is **unwound**.

*Example 1*  The usual definition of a topological **space** as a set together with a set of subsets with certain properties can be expressed in an elementary way according to definition (b) but not in a direct way according to definition (a). (But see the next remark.)

*Remark 1*  An elementary definition in the sense of (a) is also called **first order**, because the definition can be easily translated into the language of **first order logic** in a direct way. However, by incorporating the axioms of Zermelo-Fraenkel set theory into a first order theory, one can presumably state most mathematical definitions in first order logic. How this can be done is described in Chapter 7 of [Ebbinghaus, Flum and Thomas, 1984].

In spite of the fact that the ZF axioms are first order, one often hears mathematicians refer to a definition that involves quantifying over sets or over functions (as in Example 1) as non-elementary.

**Example 2**  Here is a **conceptual** definition of a left $R$-module for a ring $R$: It is an Abelian group $M$ together with a homomorphism $\phi : R \to \text{End}(M)$, where $\text{End}(M)$ denotes the ring of endomorphisms of $M$.

Now here is a more elementary definition obtained by **unwinding** the previous one: It is an Abelian group $M$ together with an operation $(r, m) \mapsto rm : R \times M \to M$ for which

a) $1m = m$ for $m \in M$, where 1 is the unit element of $R$.

b) $r(m + n) = rm + rn$ for $r \in R$, $m, n \in M$.

c) $(rs)m = r(sm)$ for $r, s \in R$, $m \in M$.

d) $(r + s)m = rm + sm$ for $r, s \in R$, $m \in M$.

One could make this a completely elementary definition by spelling out the axioms for an Abelian group. The resulting definition is elementary in both senses given above.

> The conceptual definition of left $R$-module has the advantage of making the puzzling role of "left" clear in the phrase "left $R$-module". A *right* $R$-module is a homomorphism from the *opposite ring* of $R$ to $\text{End}(M)$. This makes it apparent that the difference between left and right module is intrinsic and asymmetric, not a matter of the ostensibly symmetric and pointless distinction concerning which side you write the scalar on.
>
> On the other hand, computations on elements of the module will require knowing the laws spelled out in the elementary definition.

**Example 3**  The concept of 2-category is given both an elementary and a conceptual definition in [Barr and Wells, 1999], Section 4.8.

*Acknowledgments:* Michael Barr and Colin McLarty.

**empty set**  The **empty set** is the unique **set** with no **elements**. It is a finite set and it has **zero** elements. It is denoted by one of these symbols:

a) $\emptyset$ (a zero with a slash through it). *Citation:* (389).

b) A circle with a slash through it. *Citation:* (263).

c) An uppercase "O" with a slash through it. *Citation:* (125).

d) { }. *Citation:* (389),

e) 0 (zero) (mostly by logicians) *Citation:* (296).

f) $\phi$ (the Greek letter $\phi$ (phi)).

André Weil, who introduced this notation as part of Bourbaki, says in [Weil, 1992] that the symbol $\emptyset$ is the Norwegian letter ø, which is the letter "O" with a slash through it. (It is pronounced like the German ö.) In the Computer Modern Roman typeface of this book, it is a zero with a slash through it; a zero (0) is not as fat as an O. Both Knuth [1986] (page 128) and Schwartzman [1994] say it is a zero with a slash through it.

The notion that the symbol should be the Greek letter $\phi$ is probably a misunderstanding, but it is use by mathematicians at the blackboard quite commonly. People who use it even *call* it "phi".

See also **zero**.

***Difficulties*** Students have various difficulties with the empty set. The most basic difficulty is that they do not understand that the empty set is *something* rather than *nothing*, so that for example the set $\{\emptyset, 3, 5\}$ contains three elements, not two.

This is a perfectly natural reaction, because the basic grounding **metaphor** of (positive) **integer** is that it is the number of things in a collection, so that if you remove all the things in a collection, *you no longer have a collection.* This causes **cognitive dissonance** with the idea that the empty set is *something*. (See [Lakoff and Núñez, 2000], pages 65ff.)

Other difficulties:

- They may be puzzled by the proof that the empty set is included in every set, which is an example of **vacuous implication**.
- They also circulate a **myth** among themselves that the empty set is an *element of* every set.

"WHAT? I'M PLAYING WITH
THE EMPTY SET!"

- They may believe that the empty set may be denoted by $\{\emptyset\}$ as well as by $\emptyset$.
- They may think that the empty set is the same thing as the number 0. This may be a result of **fundamentalism**, but it also may be occasioned by the common practice among angineers and computer people of writing the number **zero** with a slash through it to distinguish it from the letter "O". I understand that some high school teachers do not allow this usage.

See **myths** and **set**.

*Acknowledgments:* Atish Bagchi

**encapsulation**    See object-process duality.

**enthymeme**    An **enthymeme** is an **argument** based partly on unexpressed beliefs. Beginners at the art of writing proofs often produce enthymemes.

***Example 1***    In the process of showing that the intersection of two equivalence relations $E$ and $E'$ is also an equivalence relation, a student may write

> "$E \cap E'$ is transitive because $E$ and $E'$ are both transitive."

This is an enthymeme; it omits stating, much less proving, that the intersection of transitive relations is transitive.

The student may "know" that it is obvious that the intersection of transitive relations is transitive, having never considered the similar question of the *union* of transitive relations, which need not be transitive.

It is very possible that the student possesses (probably subconsciously) a **malrule** to the effect that for *any* property $P$ the union or intersection of relations with property $P$ also has property $P$. The instructor very possibly suspects this. For some students, of course, the suspicion will be unjustified, but for which ones? This sort of thing is a frequent source of tension between student and instructor.

"Enthymeme" is a classical rhetorical term [Lanham, 1991]. An enthymeme is not necessarily bad, but in mathematical argument it is important that the person giving the proof know how to prove it if challenged.

*"Obvious" is the most dangerous word in mathematics.*
    *–Thomas Hobbes*

*What most experimenters take for granted before they begin their experiments is infinitely more interesting than any results to which their experiments lead.*
    *–Norbert Wiener*

**equation**   An **equation** has the form $e_1 = e_2$, where $e_1$ and $e_2$ are terms. The extensional meaning of such an equation is that $e_1$ and $e_2$ denote the same **mathematical object**. If the terms contain **free variables**, they must denote the same **variable mathematical object**.

However, the *purpose* of different equations can be utterly different, and the reader must normally depend on **context** and the perceivable structure of the equations to determine the **intensional** meaning in a particular case. The intent depends on *which part of the equation is regarded as new information*. The examples below give the commonest uses.

**Example 1**   The intent of the equation $2 \times 3 = 6$ for a grade school student may be a multiplication fact: the 6 is the new information.

**Example 2**   The intent of $6 = 2 \times 3$ may be information about a factorization: the 2 and 3 are the new information.

**Example 3**   The equation $2 \times 3 = 3 \times 2$ may be perceived as an instance of the commutative law.

**Example 4**   An equation containing **variables** may be given as a statement that the two objects are the same for all values of the variables that satisfy the hypotheses up to this point. Thus

"If $f(x)$ is constant, then $f'(x) = 0$."

This includes the case of an **identity**, as discussed under that heading. *Citations:* (85), (204).

**Example 5**   An **equation** containing **variables** may be perceived as a **constraint**. Such equations are often given by teachers and perceived by students as a command to give a simple expression for the values of the variables that make the equation true. For example, faced with $x^2 + 3x - 1 = 0$ a student should expect that the equation will have no more than two **solutions**. On the other hand, an equation such as $3x + 4y = 5$ determines a straight line: a student faced with this might be expected to give the equation of the line in slope-intercept form.

*Citations:* (43), (346).

**Example 6** An equation of the form "$y =$[expression]" may occur as the defining equation of $y$ as a function of $x$. This is discussed in the entry on function.

    *Acknowledgments:* The first three examples come from [Schoenfeld, 1985], page 66.

**equipped** Used to specify the structure attached to a set to make up a mathematical structure. Also **endowed**.

**Example 1** A semigroup is a set equipped with [endowed with] an associative binary operation. *Citations:* (68), (81).

    *Acknowledgments:* Atish Bagchi.

**equivalence relation** A **partition** $\Pi$ of a set $S$ is a set of nonempty subsets of $S$ which are pairwise disjoint and whose union is all of $S$. Here the only data are $S$ and the set $\Pi$ of subsets and the only requirements are those listed.

    An **equivalence relation** on a set $S$ is a reflexive, symmetric, transitive relation on $S$. Here the data are $S$ and the relation and the properties are those named.

    The two definitions just given provide *exactly the same class of structures.* The first one takes the set of equivalence classes as given data and the second one uses the relation as given data. Each aspect determines the other uniquely. Each definition is a different way of *presenting* the same type of structure. Thus a partition is the same thing as an equivalence relation.

    G.-C. Rota [Rota, 1997] exhibits this point of view when he says (on page 1440) "The family of all partitions of a set (also called equivalence relations) is a lattice when partitions are ordered by refinement". Literalists object to this attitude.

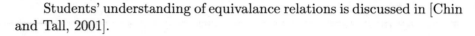

Students' understanding of equivalance relations is discussed in [Chin and Tall, 2001].

## equivalent

### 1. Equivalence of assertions

Two **assertions** are **equivalent** (sometimes **logically equivalent**) if they necessarily have the same truth values no matter how the **free variables** occurring in them are **instantiated**.

**Example 1** There are many ways to say that two assertions are equivalent. Some are listed here, all for the same assertions.

    a) A real number has a real square root **if and only if** it is nonnegative. *Citations:* (308), (354).

    b) If a real number has a real square root then it is nonnegative. Conversely, if it is nonnegative, then it has a real square root. *Citation:* (353).

    c) A real number having a real square root is equivalent to its being nonnegative. *Citation:* (67).

Other phrases are used in special cases: **in other words**, **that is**, or **equiv**alently, and **the following are equivalent**.

**Remark 1** If $P$ and $Q$ are assertions, most authors write either $P \equiv Q$ or $P \equiv Q$ to say that the two statements are equivalent. But be warned: there is a Boolean operation, often denoted by $\leftrightarrow$, with truth table

| $A$ | $B$ | $A \leftrightarrow B$ |
|---|---|---|
| T | T | T |
| T | F | F |
| F | T | F |
| F | F | T |

This is an *operation*, not a relation, and the difference between "$\leftrightarrow$" and "$\equiv$" matters. In particular, the **assertion** that three statements $P$, $Q$

and $R$ are (pairwise) equivalent is sometimes expressed by using **if and only if** or **iff** in the form "$P$ iff $Q$ iff $R$". This could be translated by "$P \equiv Q \equiv R$". Now, the connective $\leftrightarrow$ is associative, so that

$$((P \leftrightarrow Q) \leftrightarrow R) \equiv (P \leftrightarrow (Q \leftrightarrow R))$$

but the assertion "$(P \leftrightarrow Q) \leftrightarrow R$" does not have the same meaning as "$P$ iff $Q$ iff $R$" (consider the case where $P$ and $Q$ are false and $R$ is true).

In texts on discrete mathematics, [Grimaldi, 1999], [Rosen, 1991], and [Ross and Wright, 1992] all use $\leftrightarrow$ for the connective and $\equiv$ for the relation. The text [Gries and Schneider, 1993] uses $\equiv$ for the connective (and avoids the relation altogether). It appears to me that most books on logic avoid using the relation. *Citation:* (341).

**Remark 2**   One way of proving an equivalence $P \equiv Q$ is to prove that $P \Rightarrow Q$ and then that $Q \Rightarrow P$. Proving $P \Rightarrow Q$ is sometimes called the **forward direction** of the proof, and proving $Q \Rightarrow P$ is called the **converse** direction.

*Acknowledgments:* Susanna Epp, Owen Thomas.

### 2. By an equivalence relation

A phrase of the form "$x$ is equivalent to $y$" is also used to mean that $x$ and $y$ are related by an **equivalence relation**. If the equivalence relation is not clear from **context** a phrase such as "by the equivalence relation $E$" or "under $E$" may be added. *Citation:* (128).

**esilism**   This is my name for the theory espoused, usually subconsciously, by some mathematicians and logicians that the English language should be forced to mirror the notation, syntax and rules of one or another of the common forms of **first order logic**. This is a special kind of **prescriptivism**. The name is an acronym for "English Should Imitate Logic".

*Thomas Godfrey, a self-taught mathematician, was not a pleasing companion; as, like most great mathematicians I have met with, he expected universal precision in everything said, or was forever denying or distinguishing upon trifles, to the disturbance of all conversation.*

*– Benjamin Franklin*

*But all thing which that
schyneth as the gold
Ne is no gold, as
I have herd it told.*
    *−William Chaucer*

Some who read early drafts of this book
said that the sentence about gold is not
what Shakespeare wrote. Now, did I *say*
it was what Shakespeare wrote?

**Example 1**   The statement "All that glitters is not gold", translated
into logical notation the way the **syntax** indicates, gives

$$\forall x(\text{glitters}(x) \Rightarrow (\text{not gold}(x)))$$

However, its *meaning* is

$$(\text{not } \forall x(\text{glitters}(x) \Rightarrow \text{gold}(x)))$$

The "not" modifies the whole **sentence**, not the phrase "is gold".
Many, including perhaps most mathematicians, would regard this sentence as "wrong" in spite of the fact that native English speakers use
sentences like it all the time and rarely misunderstand them. Another
example is given under **order of quantifiers**.

**Remark 1**   Esilism has succeeded in ruling out the use of double negatives in educated discourse in English, but not in colloquial use in some
dialects. See [Huddleston and Pullum, 2002], Chapter 9 and the quote
from Chaucer in the sidebar. It has not succeeded in ruling out the
phenomenon described in Example 1.

To **esilize** an English sentence in the **mathematical
register** is to restate it in a form that can be mindlessly
translated into one of the usual forms of **symbolic logic**
in a way that retains the intended meaning.

**Example 2**   "Every element has an inverse" could be esilized into "For
each element $x$ there is an element $y$ that is inverse to $x$", which translates
more or less directly into $\forall x \exists y \, (\text{Inverse}(y, x))$.

**Remark 2**   A style manual for mathematical writing should address the
issue of how much esilizing is appropriate. Thus the esilizing in Example 2
is surely unnecessary, but one should avoid saying "There is an inverse
for every element", which reverses the **quantifiers**. (See Example 1 under
**order of quantifiers**.)

It may not be good style to write mathematics entirely in highly
esilized sentences, but it is quite instructive to ask students beginning

88

abstract mathematics to esilize various mathematical sentences.

*Diatribe*   Natural language has been around for thousands of years and has evolved into a wonderfully subtle tool for communication. First order logic is about a century old (although it has older precursors dating back to Aristotle) and represents an artificial form of reasoning suited to mathematics, but not to many situations in everyday life. See also literalist.

*Acknowledgments:* Susanna Epp.

**establish notation**   Mathematicians frequently say
"Let's establish some notation."
meaning they will introduce a methodical way of using certain **symbols** to refer to a particular type of **mathematical object**. This is a type of **definition** on the fly, so to speak. See also **fix** and **let**. *Citation:* (99).

**eternal**   See mathematical object.

**eureka**   See insight.

**evaluate**   To **evaluate** a function $f$ at an **argument** $x$ is to determine the value $f(x)$. See **function**. *Citation:* (288).

**every**   See universal quantifier.

**evolution**   The operation of extracting **roots** is sometimes called **evolution**. *Citation:* (337).

**example**   An **example** of a kind of **mathematical object** is a mathematical object of that kind. One also may talk about an example of a theorem; but this is often called an **illustration** and is discussed under that heading.

*The path of precept is
long, that of example
short and effectual.*

*– Seneca*

This entry provides a rough taxonomy of types of examples. The types given overlap, and whether an example is an instance of a particular type of example may depend on the circumstances (for example, the background of the reader or the student in a class).

*(a) Easy example*  An **easy example** is one that can be immediately verified with the information at hand or that is already familiar to the reader. Easy examples may be given just before or after a definition.

*Example 1*  An introduction to group theory may give as examples the integers on addition or the cyclic group of order 2, the last (I hope) presented as the group of symmetries of an isosceles triangle as well as via modular arithmetic.

*(b) Motivating example*  A **motivating example** is an example given before the definition of the concept, with salient features pointed out. Such an example gives the student something to keep in mind when reading the definition.

I have occasionally known students who object strenuously to giving an example of a concept before it is defined, on the grounds that one can't think about how it fits the definition when one doesn't know the definition. Students who feel this way are in my experience usually A students.

*Example 2*  A teacher could discuss the symmetries of the square and point out that symmetries compose and are reversible, then define "group".

*(c) Delimiting example*  A **delimiting example** (called also a **trivial example**) is one with the least possible number of elements or with **degenerate** structure.

*Example 3*  An example of a continuous function on $\mathbb{R}$ that is zero at every integer is the constant zero function. Many students fail to come up with examples of this sort ([Selden and Selden, 1998]).

*(d) Consciousness-raising example*  A **consciousness-raising example** of a kind of **mathematical object** is an example that makes the student realize that there are more possibilities for that kind of thing that

90

he or she had thought of. In particular, a consciousness-raising example may be a **counterexample** to an unconscious assumption on the student's part.

**Example 4** The function

$$f(x) = \begin{cases} x \sin \frac{1}{x} & x \neq 0 \\ 0 & x = 0 \end{cases}$$

is an example that helps the student realize that the "draw it without lifting the chalk" criterion for continuity is inadequate.

**Example 5** Example 1 of [Richter and Wardlaw, 1990] provides a diagonalizable integral matrix whose diagonal form over $\mathbb{Z}_6$ is not unique **up to** ordering. This shows that the usual assumption in diagonalization theorems that one is working over a field cannot be relaxed with impunity.

**(e) Inventory examples** Many mathematicians will check a conjecture about a type of **mathematical object** against a small number of **prototypical** examples they keep in mind when considering such objects, especially when checking conjectures. This could be called a list of **inventory examples**.

Mathematicians experienced with a concept will choose inventory examples that illustrate various possibilities. For example, $y = x^4$ is a curve with second derivative zero at a point that is not an inflection point; the dihedral group of order 8 is a nonabelian group with a proper normal subgroup that is not a direct factor.

**(f) Pathological example** A research mathematician will typically come up with a definition of a new type of **mathematical structure** based on some examples she knows about. Then further thought or conversation with colleagues will produce examples of structures that fit the definition that she had not thought of and furthermore she doesn't *want them to be the kind of thing she was thinking of.*

*Few things are harder to put up with than the annoyance of a good example.*
       *–Mark Twain*

*Turn away with fear and horror from this lamentable plague of continuous functions that do not have a derivative.*
       *–Charles Hermite*

Often the definition is modified as a result of this. Sometimes, no suitable modification seems practical and one must accept these new examples as valid. In that case they are often referred to by rude names such as **pathological** or **monster**. This was the attitude of some nineteenth-century mathematicians toward the space-filling curves, for example.

*Citations:* (39), (268).

*References:* The discussion of examples herein is drawn from [Michener, 1978] and [Bagchi and Wells, 1998a].

Occasionally an author will give a precise mathematical definition to "pathological" or "monster", for example: [Arias-De-Reyna, 1990], [Albert E. Babbitt, 1962], [McCleary, 1985]. In particular, the name "Monster group" for a particular group has become common enough that one would cause confusion by using the phrase to describe another group with pathological behavior. "Monster" in other fields is sometimes used to describe something that is merely very large in some sense.

***Difficulties*** We construct our **mental representations** of the concept primarily through examples. Experienced mathematicians know that this mental representation must always be adjusted to conform to the **definition**, but students often let examples mislead them (see **generalization**).

Generating examples is an effective way to learn a new concept. One of the marks of successful math majors is that they spontaneously generate their own examples.

*References:* [Dahlberg and Housman, 1997], [Selden and Selden, 1998].

**existential bigamy** A mistake beginning abstract mathematics students make that occurs in connection with a property $P$ of an **mathematical object** $x$ that is defined by requiring the existence of an item $y$ with a certain relationship to $x$. When students have a proof that assumes that there are two items $x$ and $x'$ with property $P$, they sometimes assume that the same $y$ serves for both of them.

***Example 1*** Let $m$ and $n$ be integers. By definition, $m$ divides $n$ if there is an integer $q$ such that $n = qm$. Suppose you are asked to prove that if $m$ divides both $n$ and $p$, then $m$ divides $n + p$. If you begin the proof by saying, "Let $n = qm$ and $p = qm \ldots$" then you are committing existential bigamy.

***Terminology***  The name is my own. The fact that Muriel and Bertha are both married (there is a person to whom Muriel is married and there is a person to whom Bertha is married) doesn't mean they are married to the same person. See **behaviors**. *Reference:* [Wells, 1995].

   *Acknowledgments:* Laurent Siebenmann.

**existential instantiation**  When $\exists(x)P(x)$ is known to be **true** (see **existential quantifier**), one may choose a **symbol** $c$ and assert $P(c)$. The symbol $c$ then denotes a **variable mathematical object** that **satisfies** $P$. That this is a legitimate practice is a standard rule of inference in **mathematical logic**. *Citation:* (189).

**existential quantifier**  For a predicate $P$, an **assertion** of the form $\exists x\, P(x)$ means that there is at least one **mathematical object** $c$ of the **type** of $x$ for which the **assertion** $P(c)$ is **true**. The **symbol** $\exists$ is pronounced "there is" or "there are" and is called the **existential quantifier**. See Remark 2 under **such that**.

***Example 1***  Let $n$ be of **type** integer and suppose $P(n)$ is the predicate "$n$ is divisible by 6". Then the **assertion** $\exists n\, P(n)$ can be expressed in the **mathematical register** in these ways:

   a) There is an integer divisible by 6. *Citations:* (261), (293).
   b) There exists an integer divisible by 6. *Citations:* (90), (247).
   c) There are integers divisible by 6. *Citations:* (195), (355).
   d) Some integers are divisible by 6. *Citation:* (292).
   e) For some integer $n$, 6 divides $n$. *Citations:* (236), (384).

***Remark 1***  If the assertion $\exists x\, P(x)$ is **true**, there may be only one **object** $c$ for which $P(c)$ is true, there may be many $c$ for which $P(c)$ is true, and in fact $P(x)$ may be true for every $x$ of the appropriate type. For example, in mathematical English the assertion, "Some of the computers have sound cards", allows as a possibility that only one computer has a sound card, and it also allows as a possibility that all the computers have sound cards. *Neither of these interpretations reflect ordinary English usage.*

In particular, in **mathematical discourse**, the **assertion**

"Some primes are less than 3."

is **true**, even though there is exactly one prime less than 3. However, I do not have an unequivocal **citation** for this. It would be a mistake to regard such a statement as false since we often find ourselves making existential statements in cases where we do not know how many **witnesses** there are.

In general, the passage from the quantifying English expressions to their interpretations as quantifiers is fraught with difficulty. Some of the basic issues are discussed in [Chierchia and McConnell-Ginet, 1990], Chapter 3; see also [Kamp and Reyle, 1993], [Gil, 1992] and [Wood and Perrett, 1997], page 12 (written for students).

See also **universal quantifier**, **order of quantifiers**, and Example 2 under **indefinite article**.

**explicit assertion**    An **assertion** not requiring **pattern recognition**.

***Example 1***    Some calculus students memorize rules in the form of explicit assertions:

"The derivative of the square of a function is 2 times the function times the derivative of the function."

A form of this rule that *does* require **pattern recognition** is:

"The derivative of $(f(x))^2$ is $2f(x)f'(x)$."

Of course, *applying* the explicit form of the rule just given requires **pattern recognition**: you must recognize that you have the square of a function and you must recognition what the function is. (This is not necessarily obvious: consider the functions $e^{2x}$, $\frac{1}{x^2}$ and $\sin^2 x$.) The point is that the rule itself is stated in a way that you don't have to decode patterns to understand what the rule says.

**Remark 1** Most definitions and theorems in mathematics do require pattern recognition and many would be difficult or impossible to formulate clearly without it.

**Remark 2** The process of converting a definition requiring pattern recognition into one that does not require it bears a striking resemblance to the way a compiler converts a mathematical expression into computer code..

**Terminology** The terminology "explicit assertion" with this meaning is my own.

**expression** See symbolic expression.

**extensional** See semantics.

**extrapolate** To assume (often incorrectly) that an **assertion** involving a certain pattern in a certain system holds for expressions of similar pattern in other systems.

**Example 1** The derivative of $x^n$ is $nx^{n-1}$, so [ERROR] the derivative of $e^x$ is $xe^{x-1}$. Of course, the patterns here are only superficially similar; but that sort of thing is precisely what causes problems for beginning abstract mathematics students.

**Example 2** The **malrule** invented by some first year calculus students that transforms $\frac{d(uv)}{dx}$ to $\frac{du}{dx}\frac{dv}{dx}$ may have been generated by **extrapolation** from the correct rule

$$\frac{d(u+v)}{dx} = \frac{du}{dx} + \frac{dv}{dx}$$

by changing addition to multiplication. The malrule

$$\sqrt{x+y} = \sqrt{x} + \sqrt{y}$$

might have been extrapolated from the distributive law

$$a(x + y) = ax + ay$$

Both these examples can be seen as the operation of a single malrule: "All operations are linear". See [Matz, 1982] and [Resnick, Cauzinille-Marmeche and Mathieu, 1987].

**factor**  If an expression $e$ is the product of several expressions $e_i$, in other words

$$e = \Pi_{i=1}^{n} e_i$$

then each $e_i$ is a **factor** of $e$. A divisor of an integer is also called a factor of the integer.

"Factor" is also used as a verb. To factor an expression is to represent it as the product of several expressions; similarly, to factor an integer (more generally an element of a structure with an associative binary operation) is to represent it as a product of integers.

See also term. *Citations:* (376), (394), (375).

**fallacy**  A **fallacy** is an error in reasoning. Two fallacies with standard names that are commonly committed by students are affirming the consequent and denying the hypothesis. See also argument by analogy and Example 1 under conditional assertion

***Terminology***  The meaning of fallacy given here is that used in this Handbook. It is widely used with a looser meaning and often connotes deliberate deception, which is not intended here.

**false symmetry**  A student may fall into the trap of thinking that some valid method or true assertion can be rearranged in some sense and still be valid or true.

The fallacy of affirming the consequent is a kind of false symmetry, and one might argue that extrapolation is another kind.  The examples below are intended to illustrate other types of false symmetry. See also **counterexample**.

I have observed all these errors in my own classes.

***Example 1*** The product of any two rational numbers is a rational number, so [ERROR] if $x$ is rational and $x = yz$ then $y$ and $z$ must be rational.

***Example 2*** If $V$ is a vector space with subspace $W$, then any basis of $W$ is included in a basis of $V$. This means [ERROR] that any basis of $V$ contains a basis of $W$ as a subset.

***Example 3*** Any subgroup of an Abelian group $A$ is normal in $A$, so [ERROR] any Abelian subgroup of a group must be normal in that group. This error may also be a case of a missing relational argument (see Behavior (iii) under **behaviors**), since being normal is a *two-place* predicate.

***Remark 1*** It would be desirable to come up with a better description of this process than "rearranged in some sense"! There may, of course, be more than one process involved.

*Acknowledgments:* Eric Schedler.

**family** A **family of sets** sometimes means an indexed set of **sets** (so differently indexed members may be the same) and sometimes merely a set of sets.

[Ross and Wright, 1992], page 686 and [Fletcher and Patty, 1988], pages 41–42 both define a family to be a set; the latter book uses "indexed family" for a tuple or sequence of sets. *Citations:* (310), (323), (358).

**field** A **field** is both a type of object in mathematical physics and a type of **object** in **abstract algebra**. The two meanings are unrelated.

**find** Used in problems in much the same way as **give**.

***Example 1*** "Find a function of $x$ whose value at 0 is positive" means "Give [an example of] a function ... "

Also used in phrases such as "we find" to mean that there is an instance of what is described after the phrase. As such, it means essentially the same thing as "there exists".

***Example 2*** "Since $\lim_{x \to \infty} f(x) = \infty$, we may find a number $x$ such that $f(x) > 10^4$."

*Citations:* (16), (77), (74), (52), (135)

**first order logic**   See mathematical logic.

**fix**   A function $f$ **fixes** a point $p$ if $f(p) = p$. This is based on this **metaphor**: you fix an object if you make it hold still (she fixed a poster to the wall). In my observation, Americans rarely use "fix" this way; in the USA, the word nearly always means "repair".

"Fix" is also used in sentences such as "In the following we fix a point $p$ one unit from the origin", which means that we will be talking about *any* point one unit from the origin (a *variable* point!) and we have **established the notation** $p$ to refer to that point. The **metaphor** behind this usage is that, because it is called $p$, every reference to $p$ is to the same value (the value is "fixed" throughout the discussion.)

*Citations:* (66), (157), (252), (280).

*Acknowledgments:* Guo Qiang Zhang.

**follow**   The statement that an **assertion** $Q$ **follows** from an assertion $P$ means that $P$ **implies** $Q$.

The word "follow" is also used to indicate that some statements after the current one are to be **grouped** with the current one, or (as in "the **following are equivalent**)" are to be grouped with each other.

*Example 1* "A set $G$ with a **binary operation** is a **group** if it satisfies the following axioms … " This statement indicates that the axioms that follow are part of the definition currently in progress.

    *Citations:* (41), (70), (129), (224).

**following are equivalent**   The phrase "**the following are equivalent**" (or "**TFAE**") is used to assert the **equivalence** of the following **assertions** (usually more than two and presented in a list). *Citation:* (307).

**for all**   See universal quantifier.

**formal**

### 1. Carefully written mathematics

Describes prose or speech that directly presents a mathematical **definition** or **argument**. In particular, a "formal proof" is a proof written in careful language with the steps made clear. This is the terminology used by Steenrod in [Steenrod *et al.*, 1975]. Similar terminology includes "formal definition". In this Handbook such formal assertions are said to be in the **mathematical register**. *Citations:* (5), (404).

### 2. Use in mathematical logic

The phrase "formal proof" is also used to mean a proof in the sense of **mathematical logic**; see **proof**. In this sense a formal proof is a **mathematical object**.

    *Reference:* [Grassman and Tremblay, 1996], pages 46–48 define formal proof as in logic.

### 3. Opposite of colloquial

The word "formal" also describes a style of writing which is elevated, the opposite of colloquial. It is not used in that meaning in this book.

**formal analogy**   A student may expect that a **notation** is to be used in a certain way by analogy with other notation based on similarity of form, whereas the **definition** of the notation requires a different use.

***Example 1***   Given real numbers $r$ and $s$ with $s$ nonzero, one can form the real number $r/s$. Given **vectors** $\vec{v}$ and $\vec{w}$, students have been known to write $\vec{v}/\vec{w}$ by **formal analogy**.

***Example 2***   In research articles in mathematics the **assertion** $A \subset B$ usually means $A$ is included as a subset in $B$. It carries no implication that $A$ is different from $B$. *Citations:* (189), (90), (338). However, the difference between "$m < n$" and "$m \leq n$" often causes students to expect that $A \subset B$ should mean $A$ is a *proper* subset of $B$ and that one should express the idea that $A$ is included in and possibly equal to $B$ by writing $A \subseteq B$. The research mathematical usage thus fails to be parallel to the usage for inequalities, which can cause **cognitive dissonance**.

   This formal analogy has resulted in a change of usage discussed further under **private language**.

***Remark 1***   I would conjecture that in Example 2, the same process is at work that is called **leveling** by linguists: that is the process that causes small children to say "goed" instead of "went".

   *Reference:* This discussion is drawn from [Bagchi and Wells, 1998a].

**formal language**   A set of **symbolic expressions** defined by a mathematical **definition**. The definition is usually given recursively.

***Example 1***   Pascal, like other modern programming languages, is a formal language. The definition, using Backus-Naur notation (a notation that allows succinct recursive definitions), may be found in [Jensen and Wirth, 1985].

***Example 2***   The languages of **mathematical logic** are formal languages. Thus **terms** and **expressions** are defined recursively on pages 14 and 15 of [Ebbinghaus, Flum and Thomas, 1984].

***Example 3*** The traditional **symbolic language** of mathematics is not a formal language; this is discussed under that entry.

See also **context-sensitive**. *Reference:* [Lewis and Papadimitriou, 1998].

## formula

*(a) Informal use* In most mathematical writing, a **formula** is an expression that allows some function to be calculated. This is analogous to the use of the word in other contexts; for example the formula for water is $H_2O$.

One might say,

"The formula for the area of a circle with radius $r$ is $\pi r^2$."

or

"The formula for the area of a circle with radius $r$ is $A = \pi r^2$."

It is not always clear whether the equation is regarded as the formula or the expression on the right side of the equation. *Citations:* (160), (57), (239).

> When teaching logic, I have frequently witnessed the difficulties students have had in remembering the difference in meaning between a formula in the sense of logic and formula as used elsewhere in mathematics. This is an example of **semantic contamination**. In this Handbook the word **assertion** is usually used instead of "formula" in the sense of logic.

*(b) In mathematical logic* In mathematical logic, a formula is **symbolic expression** in some formal language whose meaning is that of an **assertion**. It must be distinguished from **term**; for example "$x+2y$" is not a formula (it is a **term**), but "$x > y$" and $x+y = z$ are formulas. Various formalisms are described for example in [van Dalen, 1989], [Ebbinghaus, Flum and Thomas, 1984], [Mendelson, 1987] and [Hartley Rogers, 1963].

***Example 1*** The **assertion** $\forall x(x^2 \geq 0)$ could be a formula in an appropriately designed logical language.

101

**fraktur**   An alphabet formerly used for writing German that is sometimes used for mathematical **symbols**. It appears to me that its use is dying out in mathematics. Many of the forms are confusing and are mispronounced by younger mathematicians. In particular, 𝔄 may be mispronounced as "U" and 𝔍 as "T". *Citations:* (161), (183), (412).

***Remark 1***   The familiar name for this alphabet among native German speakers seems to be "Altschrift" (this is based on conversations I have had with Germans). The word "Fraktur" does occur in both German and English with this meaning, and also refers to some types of folk art.

Also called **gothic**.

| 𝔄, 𝔞  A, a | 𝔥, 𝔥  H, h | 𝔒, 𝔬  O, o | 𝔙, 𝔳  V, v |
|---|---|---|---|
| 𝔅, 𝔟  B, b | 𝔍, 𝔦  I, i | 𝔓, 𝔭  P, p | 𝔚, 𝔴  W, w |
| ℭ, 𝔠  C, c | 𝔍, 𝔧  J, j | 𝔔, 𝔮  Q, q | 𝔛, 𝔵  X, x |
| 𝔇, 𝔡  D, d | 𝔎, 𝔨  K, k | 𝔔, 𝔯  R, r | 𝔜, 𝔶  Y, y |
| 𝔈, 𝔢  E, e | 𝔏, 𝔩  L, l | 𝔖, 𝔰  S, s | ℨ, 𝔷  Z, z |
| 𝔉, 𝔣  F, f | 𝔐, 𝔪  M, m | 𝔗, 𝔱  T, t | |
| 𝔊, 𝔤  G, g | 𝔑, 𝔫  N, n | 𝔘, 𝔲  U, u | |

**free variable**   A variable in an **expression** is **free** if one can **substitute** the name of a **mathematical object** of the correct **type** for the variable and obtain the name of a (more specific but still possibly variable) mathematical object. In other words, the term determines a function with the variable as one of the arguments.

***Example 1***   The variable $x$ is free in the expression $x^2 + 1$. If you substitute 10 for $x$ you get the expression $100 + 1$ which **denotes** the **number** 101. If you substitute 10 for $x$ in $x^2 + y^2$ you get a (**variable**) **mathematical object**, $100 + y^2$.

Similarly, a **variable** in a **symbolic assertion** is **free** if it is possible to substitute the **identifier** of a specific **mathematical object** and get a meaningful statement. In particular, if one substitutes identifiers of **specific mathematical objects** for every free variable in a symbolic assertion one should get a statement that is definitely **true** or definitely false. In that sense, an assertion with free variables in it is **parametrized**; choosing values for the parameters gives a specific statement. Another way of saying this is that the assertion is a Boolean **function** of the **variables**.

**Example 2** The assertion

"$x^2 - 1 > 0$"

is not definitely true or false. However, if you substitute 2 for $x$ you get $3 > 0$ which is true, and if you substitute 0 you get a false statement.

**Remark 1** Observe that if we change the assertion in Example 2 to "$x^2 + 1 > 0$", the result is definitely true (assuming $x$ of type real) before substitution is made. Nevertheless, you can substitute a real number for $x$ in the assertion and get a statement that is definitely true or definitely false (namely definitely true), so $x$ is free. See **open sentence**.

In contrast, one cannot substitute for **bound** variables.

**Example 3** The term $\sum_{k=1}^{n} k$ becomes an expression denoting 6 if 3 is substituted for $n$. But when one substitutes a number for $k$, getting for example $\sum_{5=1}^{n} 5$, one gets nonsense; $k$ is not a free variable in the expression "$\sum_{k=1}^{n} k$", it is a **bound variable**.

**Remark 2** The preceding discussion gives a kind of behavioral definition of how free variables are used in the **mathematical register**; this definition is in the spirit of a **dictionary definition**. In texts on **mathematical logic** and on **formal languages**, freeness is generally given a recursive **mathematical definition** based on the formal recursive definition of the language. Such a recursive definition contains rules such as the following, based on a prior recursive definition of **formula**.

- $x$ is free in $x$.
- If $x$ is free in $T$ and in $T'$, then $x$ is free in $T\&T'$ (the **conjunction** of $T$ and $T'$).

- if $x$ is free in $T$, then $x$ is not free in $\exists x T$.

That sort of definition constitutes a **abstraction** of the concept of free variable defined here.

It is necessary to give such a mathematical definition of "free variable" if one is going to prove theorems about them. However, students need to know the intuition or **metaphor** underlying the concept if they are going to make fluent use of it. Most modern logic books do attempt to provide some such explanation.

*Acknowledgments:* Alonzo Church gave a similar definition of free and bound in [Church, 1942].

**function**   The concept of function in mathematics is as complex and important as any mathematical idea, though perhaps the concept of **space** has more subtleties. This long entry discusses the **syntax** we use in talking about functions, the **metaphors** behind the idea, and the difficulties connected with it.

*(a) Objects associated with a function*   When a function is discussed in the **mathematical register**, some or all of the following data will be referred to.

- An **identifier** for the function.
- The **domain** of the function.
- The **codomain** of the function
- The **argument** to the function.
- The **value** of the function at an element of the domain.
- The **graph** of the function.
- The **rule** of the function.

There is no single item in the preceding list that a discussion of a function *must* refer to. Below, I list many of the possibilities for referring to these data and the common restrictions on their use.

*(b) The identifier of a function*   (See also **value**).

*(i) Name* Functions may have **names**, for example "sine" or "the exponential function". The name in English and the **symbol** for the function may be different; for example "sine" and "sin", "exponential function" and "exp". See also **definition**.

*(ii) Local identifier* A function may be given a **local identifier**. This is by **convention** a single letter, often drawn from the Roman letters $f$ through $h$ or one of many Greek letters.

*(iii) Anonymous reference* A function may be specified without an identifier, using some form of **structural notation**. One form is to use the **defining expression** (discussed below). Other types of structural notation include **barred arrow notation** and **lambda notation**, discussed under those entries.

*(iv) Naming a function by its value at $x$* It is common to refer to a function with identifier $f$ (which may or may not be a name) as $f(x)$ (of course some other **variable** may be used instead of $x$). This is used with functions of more than one variable, too.

**Example 1** "Let $f(x)$ be a continuous function."

**Example 2** "The function $\sin x$ is bounded." *Citations:* (164), (172), (304).

*(v) The defining expression as the name of a function* It is very common to refer to a function whose **rule** is given by an **expression** $f(x)$ by simply mentioning the expression, which is called its **defining expression**. This is a special case of naming a function by its value.

**Example 3** "The derivative of $x^3$ is always nonnegative."

**Remark 1** It is quite possible that this usage should be analyzed as simply referring to the expression, rather than a function.

The **defining equation** is also used as the name of a function.

***Example 4*** The derivative of $y = x^3$ is always nonnegative.

Many authors **deprecate** this usage, but it is very common. See **writing dilemma**.

***(vi) Using the name to refer to all the values*** The name of function can be used to stand for all values. *Examples:*

"$f \geq 0$."

"$x^2$ is nonnegative."

A related phenomenon is described under **increasing function**. Compare **collective plural**. *Citation:* (335).

***(c) The argument*** The **element** of the **domain** at which the function is evaluated may be called the **argument** or the **input**. The latter word occurs most commonly with operators or algorithms. Another phrase used in some contexts is **independent variable**; then the output is the **dependent variable**. *Citations:* (93), (216), (429), (174), (362). See also **arity**.

***(d) The rule for evaluation and the graph*** For a function $f$ with domain $D$, the **graph of the function** is the set

$$\{(x, f(x) \mid x \in D\}$$

*Citation:* (33). The word is often used for the *picture* of the graph.

The **rule** for evaluation of the function is an **expression** or **algorithm** that provides a means of determining the value of the function. The rule can be a **symbolic expression** or an **algorithm**, expressed informally or in a formal language. When the rule is given by an expression $e(x)$, the definition of the function often includes the **assertion**

$$y = e(x)$$

which is called the **defining equation** of the function.

**Example 5** "Consider the function given by $y = x^2 + 1$." The defining equation is "$y = x^2 + 1$".

**Example 6** "Consider the function $f(x) = x^2 + 1$." Note that this gives the defining equation as a **parenthetic assertion**. In this expression the **variable** $x$ is bound (see Example 3 under **bound variable**).

**Remark 2** A subtle point, which perhaps students should not be bothered with too early, is that a function will always have a graph but it need not always have a defining rule. This is because the number of possible rules (which are strings in a finite alphabet) is only countably infinite, but the cardinality of the set of all functions between two infinite sets is uncountable. *Citations:* (28), (45), (63), (302), (381), (396).

Because of the practice of using defining equations, students often *regard* a function as an equation [Thompson, 1994], pp 24ff. So do teachers [Norman, 1992].

*Disjunctive definitions* An expression involving disjunctions can confuse students, who don't recognize it as one expression defining one function.

**Example 7** "Let $f(x) = \begin{cases} x+1 & \text{if } x > 2 \\ 2x - 1 & \text{otherwise.} \end{cases}$" *Citations:* (140), (281).

*(e) How one thinks of functions* A mathematician's **mental representation** of a function is generally quite rich and involves several different **metaphors**. Some of the more common ways are noted here. These points of view have blurry edges!

*(i) Expression to evaluate* Function as **expression** to evaluate. This is the motivation for item (v) under "The identifier of a function" earlier in this entry. It is the image behind statements such as "the deriva-

It appears to me that some mathematicians avoid using the word "function" for functions that do not act on numbers, perhaps for reasons of readability. Instead, they use words such as **functional**, **operator**, or **operation**. I have heard secondhand stories of mathematicians who objected to using the word "function" for a **binary operation** such as addition on the integers, but I have never seen that attitude expressed in print. In this text functions are *not* restricted to operating on numbers. See also **mapping**.

tive of $x^3$ is $3x^2$". This gives an intensional semantics to the expression. *Citation:* (93).

*(ii) Graph* Function as graph. A function illustrated graphically is usually numerical. This provides a *picture* of the function as a *relation* between argument and value; of course it is a special kind of relation. *Citation:* (84).

*(iii) Dependency relation* Function as a **dependency relation**. This is the metaphor behind such descriptions as "let $x$ depend smoothly on $t$". It is related to the graph point of view, but may not carry an explicit picture; indeed, an explicit picture may be impossible. *Citation:* (239).

*(iv) Transformer* Function as **transformer**, that takes an object and turns it into another object. In this picture, the function $F(x) = x^3$ transforms 2 into 8. This is often explicitly expressed as a "**black box**" interpretation, meaning that all that matters is input and output and not how it is performed. This point of view is revealed by such language as "2 becomes 8 under $f$".

*(v) Algorithm* Function as algorithm or set of rules that tell you how to take an input and convert it into an output. This is a metaphor related to those of function as expression and as transformer, but the actual process is *implicit* in the expression view (in the intensional semantics of the expression) and *hidden* in the transformer (black box) view. *Citation:* (385).

*(vi) Relocator* Function as **relocator**. In this version, the function $F(x) = x^3$ moves the point at 2 over to the location labeled 8. This is the "alibi" interpretation of [Mac Lane and Birkhoff, 1993] (page 256). It is revealed by such language as "$f$ takes 2 to 8". *Citation:* (423).

*(vii) Map* Function as map. This is one of the most powerful metaphors in mathematics. It takes the point of view that the function

$F(x) = x^3$ renames the point labeled 2 as 8. A clearer picture of a function as a map is given by some function that maps the unit circle onto, say, an ellipse in the plane. The ellipse is a map of the unit circle in the same way that a map of Ohio has a point corresponding to each point in the actual state of Ohio (and preserving shapes in some approximate way). This is something like the "alias" interpretation of [Mac Lane and Birkhoff, 1993]: The point on the map labeled "Oberlin", for example, has been *renamed* "Oberlin". *Citations:* (252), (357).

*References:* [Lakoff and Núñez, 1997], [Selden and Selden, 1992]

*(f) Mathematical definitions of function*   Texts in calculus and discrete mathematics often **define** the concept of function as a **mathematical object**. There are two nonequivalent definitions in common use.

a) A function is a set of ordered pairs with the **functional property**: pairs with the same first coordinate have the same second coordinate.

b) A function consists of two sets called the **domain** and the **codomain** of the function, and a set of ordered pairs with the functional property, subject to the requirements that

- the domain must be exactly the set of first coordinates of the graph, and
- the codomain must include all the second coordinates.

The set of ordered pairs in the second definition is called the **graph** of the function.

**Example 8**   Consider two sets $A$ and $B$ with $A \subseteq B$, and consider the identity function from $A$ to $A$ and the inclusion function $i : A \to B$ defined by $i(a) = a$.

Under the definition (a) above, the identity function and the inclusion function are the same function. Under the definition (b), they are different functions because they have different codomains, even though they have the same **graph**.

Authors vary much more in the treatment of the codomain than they do in the treatment of the **domain**. Many authors use definition (a) and do not mention a codomain at all. Others don't make it clear whether they require it or not. Even when authors do require specification of the codomain, the specification is often an empty gesture since the text fudges the question of whether two functions with the same **domains** and same **graphs** but different codomains are really different.

See **composite**, **range** and **image**. *Citations:* (65), (180).

**(g) Difficulties** Typically, the definition of "function" does not correspond very well with actual usage. For example, one generally does not see the function expressed in terms of ordered pairs, one more commonly uses the $f(x)$ **notation** instead. To avoid this discrepancy, I suggested in [Wells, 1995] the use of a **specification** for functions instead of a definition. I have discussed the discrepancy in the treatment of the codomain in the preceding section.

These discrepancies probably cause some difficulty for students, but for the most part students' difficulties are related to their inability to **reify** the concept of function or to their insistence on maintaining just one **mental representation** of a function (for example as a set of ordered pairs, a graph, an **expression** or a **defining equation**).

There is a large literature on the difficulties functions cause students, I am particularly impressed with [Thompson, 1994]. Another important source is the book [Harel and Dubinsky, 1992] and the references therein, especially [Dubinsky and Harel, 1992], [Norman, 1992], [Selden and Selden, 1992], [Sfard, 1992]. See also [Vinner and Dreyfus, 1989], [Eisenberg, 1992] and [Carlson, 1998]. [Hersh, 1997a] discusses the confusing nature of the word "function" itself.

*Acknowledgments:* Michael Barr.

**functional**  The word **functional** is used as a noun to denote some special class of functions. The most common use seems to be to denote a function whose domain consists of **vectors**, **functions**, or elements of some abstract **space** such as a normed linear space whose **motivating examples** are function spaces. But not all uses fit this classification. *Citations:* (45), (216), (247), (299).

**functional knowledge**  Functional knowledge is "knowing *how*"; **explicit knowledge** is "knowing *that*". Functional knowledge is also called **implicit knowledge**.

*Example 1*  Many people, upon hearing a simple tune, can sing it accurately. If two successive notes differ by a whole step, as for examples the second and third notes of "Happy Birthday", they will sing the sequence correctly. They have the functional knowledge to be able to copy a simple enough tune. Some with musical training will be able to say that this jump is a whole step, even though they had never thought about the tune before; they are able to turn their functional knowledge into **explicit knowledge**.

> "Whole step" is a technical term in music theory for the relation of the frequencies of the two notes.

*Example 2*  Consider two students Hermione and Lucy in a first course in number theory. The instructor sets out to prove that if the square of an integer is odd, then the integer is odd. Her proof goes like this:

"Suppose the integer $n$ is even. Then $n = 2k$ for some integer $k$, so $n^2 = 4k^2$ Thus $n^2$ is even as well. QED."

Hermione immediately understands this proof and is able to produce similar proofs using the **contrapositive** even though she has never heard of the word "contrapositive". Lucy is totally lost; the proof makes no sense at all: "Why are you assuming it is even when you are trying to prove it is odd?" (I suspect some students are so stumped they don't even formulate a question like this.)

111

In first classes involving rigorous proofs, *this sort of thing happens all the time.* Clearly, Hermione has some sort of functional knowledge that *enables her to grasp the logical structure of arguments* without any explicit explanation. And Lucy does not have this functional knowledge; she needs tutoring in mathematical reasoning.

**Remark 1**   Some students appear to me to have a natural talent for acquiring the ability to grasp the logical structure of arguments, without being explicitly taught anything about the **translation problem** or **mathematical logic**. Indeed, many professional mathematicians know very little of the terminology of logic but have no trouble with understanding the logical structure of **narrative arguments**. Other students have difficulty even when taught about the logical structure explicitly.

This may be the fundamental difference between those who have a **mathematical mind** and those who don't. It is likely that (a) most students can learn mathematics at the high school level even though (b) some students have special talent in mathematics.

The situation seems to be similar to that in music. Music educators generally believe that most people can be taught to sing in time and on key. Nevertheless, it is obvious that some children have special musical talent. The parallel with mathematics is striking.

**fuzzy**   See concept.

**generalization**

**(a) Legitimate generalization**   To generalize a mathematical concept $C$ is to find a concept $C'$ with the property that instances of $C$ are also instances of $C'$.

**(i) Expansive generalization**   One may generalize a concept by changing a datum of $C$ to a **parameter**. This is **expansive generalization**.

112

***Example 1***   $\mathbb{R}^n$, for arbitrary positive integer $n$, is a generalization of $\mathbb{R}^2$. One replaces the ordered pairs in $\mathbb{R}^2$ by ordered $n$-tuples, and much of the spatial structure (except for the **representation** of $\mathbb{R}^2$ using complex numbers) and even some of our **intuitions** carry over to the more general case.

    ***(ii) Reconstructive generalization***   Some generalizations may require a substantial cognitive reconstruction of the concept. This is **reconstructive generalization**.

***Example 2***   The relation of the concept of abstract real vector space to $R^n$ is an example of a reconstructive generalization. One forgets that the elements are $n$-tuples and adopts axioms on a set of points to make a real vector space. Of course, one can do an expansive generalization on the field as well, changing $\mathbb{R}$ to an arbitrary field.

    Another example occurs under **continuous**.

***Remark 1***   The suspicious reader will realize that I have finessed something in this discussion of vector spaces. If you have only a naive idea of the real plane as a set of points, then before you can make an expansive generalization to $\mathbb{R}^n$ you must *reconstruct* the real plane by **identifying** each point with its pair of coordinates. It appears to me that this reconstruction happens to some students in high school and to others in college.

***Remark 2***   The relation between reconstructive generalization and **abstraction** should be studied further.

    The names "expansive" and "reconstructive" are due to [Tall, 1992a].

**(b) Generalization from examples**   The idea of generalization discussed above is part of the legitimate methodology of mathematics. There is another process often called generalization, namely **generalization from**

> It appears to me that the usual meaning of the word "generalization" in colloquial English is generalization from examples. Indeed, in colloquial English the word is often used in a derogatory way. The contrast between this usage and the way it is used in mathematics may be a source of **cognitive dissonance**.

**examples**. This process is a special case of **extrapolation** and is a common method of reasoning that however can lead to incorrect results in mathematics.

***Example 3***    All the limits of sequences a student knows may have the property that the limit is not equal to any of the terms in the sequences, so the student generalizes this behavior with the **myth** "A sequence gets close to its limit but never equals it". Further discussion of this is in Example 1 under **limit**. See also **extrapolate**, **myths** and **syntax**.

**generic**    See **mathematical object**.

**give**    "Give" is used in several ways in the mathematical register, often with the same sense it would be used in any academic text ("we give a proof ... ", "we give a construction ... "). One particular mathematical usage: to **give** a **object** means to describe it sufficiently that it is **uniquely** determined. Thus a phrase of the form "give an $X$ such that $P$" means describe a **object** of type $X$ that satisfies **predicate** $P$. The description may be by providing a **determinate identifier** or it may be a definition of the object in the **mathematical register**.

***Example 1***    "Problem: Give a function of $x$ that is positive at $x = 0$." A correct answer to this problem could be "the cosine function" (provide an identifier), or "the function $f(x) = x^2+1$" (in the calculus book dialect of the mathematical register).

"Given" may be used to introduce an expression that defines an object.

***Example 2***    One could provide an answer for the problem in the preceding example by saying:

"the function $f : \mathbb{R} \to \mathbb{R}$ **given** by $f(x) = 2x + 1$."

The form **given** is also used like **if**.

**Example 3**  "Given sets $S$ and $T$, the **intersection** $S \cap T$ is the set of all objects that are elements of both $S$ and $T$."

See also find. *Citations:* (78), (126), (152), (219), (259), (391).

**global**  See local.

**global identifier**  A **global identifier** in a mathematical text is an identifier that has the same meaning throughout the text. An identifier defined only in a section or paragraph is a local identifier.

Global identifiers may be classified into three types:

*(i) Global to all of mathematics*  Some global identifiers are used by nearly all authors, for example "=", mostly without definition. Some global identifiers such as $\pi$ and $e$ are sometimes overridden in a particular text. Even "=" is sometimes overridden; for example, one may define the rationals as equivalence classes of ordered pairs of integers, and say we write $a/b$ for $(a, b)$ and $a/b = a'/b'$ if $(a, b)$ is equivalent to $(a', b')$.

*(ii) Global to a field*  Some are used by essentially all authors in a given field and generally are defined only in the most elementary texts of that field.

**Example 1**  The integral sign is global to any field that uses the calculus. This seems never to be overridden in the context of calculus, but it does have other meanings in certain special fields (ends and coends in category theory, for example).

*(iii) Global to a text*  A global identifier may be particular to a given book or article and defined at the beginning of that text.

---

Global identifiers specific to a given text impose a burden on the memory that makes the text more difficult to read, especially for grasshoppers. It helps to provide a glossary or list of symbols, and to use type labeling. Steenrod [1975] says global symbols specific to a text should be limited to five. Mnemonic global identifiers of course put less burden on the reader.

---

**Remark 1**  The classification just given is in fact an arbitrary division into three parts of a continuum of possibilities.

*Acknowledgments:* Thanks to Michael Barr, who made valuable suggestions concerning an earlier version of this entry.

**gothic**   The German fraktur alphabet is sometimes called gothic, as is an alphabet similar to fraktur but easier to read that is used as newspaper titles.

Certain sans-serif typefaces are also called gothic.

**graph**   The word "graph" has two unrelated meanings in undergraduate mathematics:

a) The graph of a function.

b) A structure consisting of nodes with directed or undirected edges that connect the nodes is called a (directed or undirected) **graph**. The actual mathematical definitions in the literature vary a bit. *Citation:* (422).

Moreover, in both cases the word "graph" may also be used for *drawings* of (often only part of) the **mathematical objects** just described. *Citations:* (102), (100), (22).

**grasshopper**   A reader who starts reading a book or article at the point where it discusses what he or she is interested in, then jumps back and forth through the text finding information about the ideas she meets. Contrasted with someone who starts at the beginning and reads straight through.

***Terminology***   The terminology is due to Steenrod [1975]. Steenrod calls the reader who starts at the beginning and reads straight through a **normal** reader, a name which this particular grasshopper resents.

**Greek alphabet**   Every letter of the Greek alphabet except omicron (O,$o$) is used in mathematics. All the **lowercase** forms and all those **uppercase** forms that are not identical with the Roman alphabet are used. Students and young mathematicians very commonly mispronounce element 79 some of them. The letters are listed here with pronunciations and with some comments on usage. Some information about the common uses of many of these letters is given in [Schwartzman, 1994].

**Pronunciation key:** ăt, āte, bĕt, ēve, pĭt, rīde, cŏt, gō, fo͞od, fo͝ot, bŭt, mūte, ə (schwa) the neutral unaccented vowel as in ago (əgō) or focus (fōkəs). A prime after a syllable indicates primary accent; double prime secondary accent, as in secretary (sĕ′krətă″rĭ) (American pronunciation). (Br) indicates that the pronunciation is used chiefly in countries whose education system derived from the British system (including Britain, Australia, New Zealand, South Africa).

Most Greek letters are pronounced differently in modern Greek; $\beta$ for example is pronounced vē′ta (last vowel as "a" in father).

A, $\alpha$   Alpha, ăl′fə. *Citation:* (50).

B, $\beta$   Beta, bā′tə or bē′tə (Br). *Citation:* (50).

Γ, $\gamma$   Gamma, gă′mə. *Citations:* (107), (118), (138).

Δ, $\delta$   Delta, dĕl′tə. *Citations:* (26), (48).

E, $\epsilon$ or $\varepsilon$   Epsilon, ĕp′sələn, ĕp′səlŏn″, or ĕpsī′lən. Note that the symbol $\in$ for **elementhood** is not an epsilon. *Citations:* (103), (395).

Z, $\zeta$   Zeta, zā′tə or zē′tə (Br).. *Citation:* (375).

H, $\eta$   Eta, ā′tə or ē′tə (Br). *Citation:* (129).

Θ, $\theta$ or $\vartheta$   Theta, thā′tə or thē′tə (Br). *Citations:* (211), (184), (395).

I, $\iota$   Iota, īō′tə. *Citation:* (310).

K, $\kappa$   Kappa, kăp′ə. *Citation:* (88).

Λ, $\lambda$   Lambda, lăm′də. *Citation:* (88).

M, $\mu$   Mu, mū. *Citation:* (249).

N, $\nu$   Nu, no͞o or nū. *Citation:* (157).

Ξ, ξ   Xi. I have heard ksē, sī and zī. Note that the pronunciation sī is also used for $\psi$ (discussed further there). *Citations:* (106), (301).

O, $o$   Omicron, ŏ′mĭkrŏn″ or ō′mĭkrŏn″.

Π, π   Pi, pī. To the consternation of some students beginning abstract mathematics, $\pi$ is very commonly used to mean all sorts of things besides the ratio of the circumference of a circle to its diameter. *Citations:* (184), (257), (320).

P, ρ   Rho, rō. *Citations:* (322), (379).

Σ, σ   Sigma, sĭg′mə. *Citations:* (423), (101), (293), (397).

T, τ   Tau, pronounced to rhyme with cow or caw. *Citation:* (376).

Υ, υ   Upsilon. The first syllable can be pronounced ōōp or ŭp and the last like the last syllable of epsilon. *Citations:* (344), (431).

Φ, φ or φ   Phi, fī or fē. For comments on the symbol for the empty set, see **empty set**. *Citations:* (257), (317).

I have heard two different young mathematicians give lectures containing both $\phi$ and $\psi$ who pronounced one of them fī and the other fē. I have also heard lecturers pronounce both letters in exactly the same way.

X, χ   Chi, pronounced kī. I have never heard anyone say kē while speaking English (that would be the expected vowel sound in European languages). German speakers may pronounce the first consonant like the ch in "Bach". *Citation:* (379).

Ψ, ψ   Psi, pronounced sī, sē, psī or psē. *Citations:* (322), (395), (397).

Ω, ω   Omega, ōmā′gə or ōmē′gə. *Citations:* (14), (132), (289), (397).

*Acknowledgments:* Gary Tee.

**grouping**   Various syntactical devices are used to indicate that several **assertions** in the **mathematical register** belong together as one logical unit (usually as a **definition** or **theorem**). In the **symbolic language** this is accomplished by **delimiters**. In general mathematical prose various devices are used. The statement may be **delineated** or **labeled**, or phrases from the general academic register such as "the following" may be used. Examples are given under **delimiter** and **follow**.

**guessing**   If the **definition** of a **mathematical object** determines the object **uniquely**, then **guessing** at the answer to a problem and then using the definition or a theorem to prove it is correct is legitimate, but some students don't believe this.

***Example 1***   It is perfectly appropriate to guess at an antiderivative and then prove that it is correct by differentiating it. Many students become uncomfortable if a professor does that in class.

This attitude is a special case of **algorithm addiction**.

**hanging theorem**   A **theorem** stated at the point where its proof is completed, in contrast to the more usual practice of stating the theorem and then giving the proof. *References:* The name is due to Halmos [Steenrod *et al.*, 1975], page 34, who **deprecates** the practice, as does [Krantz, 1997], page 68.

**hat**   See **circumflex**.

**hidden curriculum**   **Covert curriculum**.

**hold**   An **assertion** $P$ about **mathematical objects** of type $X$ **holds** for an instance $i$ of $X$ if $P$ becomes **true** when $P$ is **instantiated** at $i$.

***Example 1***   Let the type of $x$ be real and let $P$ be the **assertion**

$$f(x) > -1$$

Then $P$ holds when $f$ is instantiated as the sine function and $x$ is instantiated as 0. Typical usage in the **mathematical register** would be something like this: "$P$ holds for $f = \sin$ and $x = 0$."

"Hold" is perhaps most often used when the instance $i$ is bound by a quantifier.

***Example 2***   "$x^2 + 1 > 0$ holds for all $x$."
    *Citation:* (28), (129), (363).

**hyphen**   A hyphen is sometimes used to indicate the **variable** in the definition of a **function**, especially when the function is defined by holding some of the variables in a multivariable function fixed.

***Example 1***   Let $f(x, y)$ be a function. For a fixed $a$, let $f(-, a)$ be the corresponding function of one variable.

    *Citation:* (44).

**I**   The symbol I has several common uses, often without explanation:
- I may be used to denote the unit **interval**, the set of real numbers $x$ for which $0 \leq x \leq 1$. It may however denote *any* bounded interval of real numbers.
- I may be used as the name of an arbitrary index set.
- For some authors, I or $\mathbb{I}$ is defined to be the set of integers; however, $\mathbb{Z}$ seems to be more common in this respect.
- $I$ may be used to denote the identity function.
- $I$ is also used to denote the identity matrix for a dimension given by context.

*Citations:* (28), (40), (68), (97); (252), (310).

**i.e.**   The expression "i.e." means **that is**. It is very commonly confused with "**e.g.**", meaning "for example", by students and sometimes by professors. This confusion sometimes gets into the research literature. Using i.e. and e.g. should be **deprecated**. *Citations:* (67), (90), (364), (115), (159).

**identifier**   An **identifier** is a **name** or **symbol** used as the name of a **mathematical object**. Symbols and **names** are defined in their own entries; each of these words has precise meanings in this Handbook that do not coincide with common use. In particular, a symbol may consist of more than one **character** and a name may be a word or a phrase.

We discuss the distinction between name and symbol here. A **name** is an English **noun phrase**. A **symbol** is a part of the **symbolic language** of mathematics.

**Example 1** The **expressions** $i$, $\pi$ and sin can be used in symbolic expressions and so are symbols for certain **objects**. The phrase "the sine function" is a name. If a **citation** is found for "sine" used in a symbolic expression, such as "sine$(\pi)$", then for that author, "sine" is a symbol.

**Remark 1** The number $\pi$ does not appear to have a nonsymbolic **name** in common use; it is normally identified by its **symbol** in both English **discourse** and **symbolic expressions**. The complex number $i$ is also commonly referred to by its symbol, but it can also be called the **imaginary unit**. *Citations:* (185), (332).

**Remark 2** I have not found examples of an identifier that is not clearly either a **name** or a **symbol**. The **symbolic language** and the English it is embedded in seem to be quite sharply distinguished.

**Terminology** I have adopted the distinction between name and symbol from [Beccari, 1997], who presumably is following the usage of [ISO, 1982] which at this writing I have not seen yet.

**identify** To **identify** an **object** $A$ with another object $B$ is to regard them as identically the same object. This may be done via some formalism such as an amalgamated product or a pushout in the sense of category theory, but it may also be done in a way that suppresses the formalism (as in Example 1 below).

**Example 1** The Möbius strip may be constructed by identifying the edge

$$\{(0, y) \mid 0 \le y \le 1\}$$

of the unit square with the edge

$$\{(1,y) \mid 0 \leq y \leq 1\}$$

in such a way that $(0,y)$ is identified with $(1, 1-y)$.

There is another example in Remark 1 under **generalization**.

***Remark 1*** One may talk about identifying one *structure* (space) with another, or about identifying individual *elements* of one structure with another. The word is used both ways as the citations illustrate. Example 1 uses the word both ways in the same construction.

***Remark 2*** One often identifies objects without any formal construction and even without comment. That is an example of **conceptual blending**; more examples are given there. *Citations:* (186), (285), (399).

***Remark 3*** In ordinary English, "identify" means to give a name to. This presumably could cause **cognitive dissonance** but I have never observed that happening myself.

**identity**    This word has three common meanings.

### 1. Equation that always holds

An **identity** in this sense is an **equation** that holds between two **expressions** for any valid values of the variables in the expressions. Thus, for real numbers, the equation $(x+1)^2 = x^2 + 2x + 1$ is an identity. But in the **assertion**

"If $x = 1$, then $x^2 = x$,"

the equation $x^2 = x$ would not be called an identity. The difference is that the equation is an identity if the only restrictions imposed on the variables are one of **type**. This is a psychological difference, not a mathematical one. *Citation:* (36),

Sometimes in the case of an identity the symbol $\equiv$ is used instead of the equals sign. *Citation:* (37).

### 2. Identity element of an algebraic structure

If $x \, \Delta \, e = e \, \Delta \, x$ for all $x$ in an algebraic structure with **binary operation** $\Delta$, then $e$ is an **identity** or **identity element** for the structure. Such an element is also called a **unit element** or **unity**. This can cause confusion in ring theory, where a **unit** is an invertible element. *Citations:* (193), (251).

### 3. Identity function

For a given set $S$, the function from $S$ to $S$ that takes every element of $S$ to itself is called the **identity function**. This is an example of a **polymorphic** definition. *Citation:* (134).

# if

**(a) Introduces conditional assertion** The many ways in which "if" is used in translating conditional assertions are discussed under **conditional assertion**.

**(b) In definitions** It is a **convention** that the word if used to introduce the **definiens** in a **definition** means "if and only if".

**Example 1** "An integer is **even** if it is divisible by 2." *Citation:* (); (261). Some authors regularly use "if and only if" or "iff". *Citations:* (126), (377). This is discussed (with varying recommendations) in [Gillman, 1987], page 14; [Higham, 1993], page 16; [Krantz, 1997], page 71; [Bagchi and Wells, 1998b].

**Remark 1** It is worth pointing out (following [Bruyr, 1970]) that using "if and only if" does not rid the definition of its special status: there is still a **convention** in use. In the definition of "even" above, the left side (the **definiendum**) of the definition, "An integer is **even**", does not attain the status of an **assertion** until the whole definition is read. In an ordinary **statement** of equivalence, for example "An integer is divisible by 2 if and only if it is divisible by $-2$", both sides are assertions to begin with. In a definition there is a special convention in effect that allows one to use a syntax that treats the definiendum as an assertion even though before the definition takes place it is meaningless.

Because of this, insisting on "if and only if" instead of "if" in definitions is not really an example of **esilism** because "if and only if" is not used in definitions in first order logic. The usual formalisms for first order logic use other syntactic devices, probably because of the phenomenon just described.

*(c) In the precondition of a definition*    "If" can be used in the **precondition** of a **definition** to introduce the **structures** necessary to make the definition, in much the same way as **let**. See Example 5 under **definition**. *Citation:* (318).

See also the discussion under **let**.

**if and only if**    This phrase denotes the relation **equivalent** that may hold between two assertions. See **context-sensitive** and **if**.

This phrase may be abbreviated by **iff**. *Citations:* (63), (110), (253).

**illustration**    A drawing or computer rendering of a curve or surface may be referred to as an **illustration**. Thus a drawing of (part of) the graph of the equation $y = x^2$ would be called an illustration. The word is also used to refer to an instance of an **object** that satisfies the hypotheses and conclusion of a **theorem**. (This is also called an **example** of the theorem.)

*Example 1*    A professor could illustrate the theorem that a function is **increasing** where its derivative is positive by referring to a drawing of the graph of $y = x^2$.

*Example 2*    The fact that subgroups of an Abelian **group** are normal could be illustrated by calculating the cosets of the two-element subgroup of $Z_6$. This calculation might not involve a picture or drawing but it could still be called an illustration of the theorem. *Citations:* (50), (70).

**image**   The word **image**, or the phrase **concept image**, is used in mathematical education to refer to what is here called the **mental representation** of a concept.

    This word also has a **polymorphic** mathematical meaning (the image of a function) discussed under **overloaded notation**.

**implication**   See conditional assertion.

**implicit knowledge**   See functional knowledge.

**imply**   See conditional assertion.

**in**   In is used in **mathematical discourse** in all its English meanings, as well as in some meanings that are peculiar to mathematics.

a) $A$ is in $B$ can mean $A \in B$. *Citation:* (226).

b) $A$ is in $B$ can mean $A \subseteq B$. *Citation:* (322).

c) One may say $A$ is in $B$ when $A$ is an equation whose solution set is included in $B$, or a geometric figure whose points are included in $B$, or a sequence whose entries are in $B$. For example,
> "The unit circle $x^2 + y^2 = 1$ is in the Euclidean plane."

*Citations:* (28), (213), (265), (359), (375), (384).

d) One may say $A$ is in $B$ when $A$ is in the expression $B$ as some syntactic substructure. For example, $x$ is a variable in $3x^2 + 2xy^3$, and $3x^2$ is a term in $3x^2 + 2xy^3$. *Citation:* (322),

e) $A$ is $P$ in $B$, where $P$ is a property, may mean that $A$ has property $P$ with respect to $B$, where $B$ is a constituent of $A$ or a related **structure** (for example a containing structure). Thus one talks about $A$ being normal in $B$, where $A$ is a subgroup of the group $B$. As another example, $3x^2 + 2xy^3$ is differentiable in $y$ (and $x$). (122), (122), (178), (205), (224), (276), (408).

f) One may describe an intersection using "in". For example, The sets $\{1, 2, 3, 4\}$ and $\{1, 3, 5, 7\}$ intersect in $\{1, 3\}$ (or intersect in 1 and 3.) *Citation:* (369).

g) In the phrase "in finite terms". *Citation:* (293).

**in general**   The phrase "in general" occurs in at least two ways in mathematical statements. (One may often use "generally" with the same meaning.)

***Example 1***   "The equation $x^2 - 1 = (x - 1)(x + 1)$ is true in general." *Citation:* (154)

***Example 2***   "In general, not every sub**group** of a group is normal." *Citations:* (83), (215), (243)

Example 1 asserts that the equation in question is always true. Example 2 does not make the analogous claim, which would be that *no* subgroup of a group is normal. These two examples illustrate a pattern: "In general, $P$" tends to mean that $P$ is always true, whereas "In general, not $P$" means that $P$ is not necessarily true.

*Acknowledgments:* Owen Thomas.

**in other words**   This phrase means that what follows is **equivalent** to what precedes. Usually used when the equivalence is easy to see. *Citation:* (273).

**in particular**   Used to specify that the following statement is an in-**stantiation** of the preceding statement, or more generally a consequence of some of the preceding statements. The following statement may indeed be equivalent to the preceding one, although that flies in the face of the usual meaning of "particular".

***Example 1***   "We now know that $f$ is differentiable. In particular, it is continuous." *Citations:* (114), (283), (430).

**in terms of**   See term.

**in your own words**   Students are encouraged in high school to describe things "in your own words". When they do this in mathematics class, the resulting reworded definition or theorem can be seriously misleading or wrong. It might be reasonable for a teacher to encourage students to rewrite mathematical statements in their own words and then *submit them to the teacher*, who would scrutinize them for dysfunctionality.

***Example 1***   Students frequently use the word **unique** inappropriately. A notorious example concerns the definition of **function** and the definition of **injective**, both of which students may reword *using the same words*:

> "A function is a relation where there is a unique output for every input."

> "An injective function is one where there is a unique output for every input."

See also **continuous**.

**include**   For sets $A$ and $B$, $B$ **includes** $A$, written $A \subseteq B$ or $A \subset B$, if every **element** of $A$ is an element of $B$. See the discussions under **contain** and **formal analogy**. *Citations:* (122), (195), (262), (338).

**increasing**   An **increasing function** is a function $f$ **defined on** some ordered set $S$ with the **property** that if $x < y$ then $f(x) < f(y)$. A sequence $a_0, a_1, \ldots$ is **increasing** if it is true that $i < j$ implies $a_i < a_j$.

A similar phenomenon occurs with **decreasing**, **nondecreasing** and **nonincreasing**. This is related to the phenomenon described under "Using the name to refer to all the values" under **function**. See also **time**. *Citations:* (233), (394)

**indefinite article**   The word "a" or "an" is the **indefinite article**, one of two **articles** in English.

**(a) *Generic use***   In mathematical writing, the indefinite article may be used in the subject of a clause with an **identifier** of a type of **mathematical object** (producing an **indefinite description**) to indicate an arbitrary object of that type. Note that plural indefinite descriptions do not use an article. This usage occurs outside mathematics as well and is given a theoretical treatment in [Kamp and Reyle, 1993], section 3.7.4.

**Example 1**
   "Show that an integer that is divisible by four is divisible by two."

*Correct interpretation:* Show that *every* integer that is divisible by four is divisible by two. *Incorrect interpretation:* Show that some integer that is divisible by four is divisible by two. Thus in a sentence like this it the indefinite article has the force of a **universal quantifier**. Unfortunately, this is also true of the **definite article** in some circumstances; more examples are given in the entry on **universal quantifier**. *Citations:* (156), (179), (353) (for the indefinite article); (354) (for the definite article).

**Remark 1**   This usage is **deprecated** by Gillman [1987], page 7. Hersh [1997a] makes the point that if a student is asked the question above on an exam and answers, "24 is divisible by 4, and it is divisible by 2", the student should realize that with that interpretation the problem is too trivial to be on the exam.

**Remark 2**   An **indefinite description** apparently has the force of universal quantification only in the subject of the clause. Consider:

128

a) "A number divisible by 4 is even." (Subject of sentence.)

b) "Show that a number divisible by 4 is even." (Subject of subordinate clause.)

c) "Problem: Find a number divisible by 4." (Object of verb.) This does not mean find *every* number divisible by 4; one will do.

**Remark 3**  In ordinary English sentences, such as

"A wolf takes a mate for life."

([Kamp and Reyle, 1993], page 294), the meaning is that the **assertion** is true for a *typical* individual (typical wolf in this case). In mathematics, however, the assertion is required to be true without exception. See **concept** and **prototype**.

*(b) Existential meaning*  An indefinite description may have **existential** force.

**Example 2**  "A prime larger than 100 was found in 2700 B.C. by Argh P. Ugh." This does *not* mean that Mr. Ugh found *every* prime larger than 100. In this case the indefinite description is the subject of a *passive* verb, but in ordinary English indefinite subjects of active verbs can have existential force, too, as in "A man came to the door last night selling toothbrushes". I have found it difficult to come up with an analogous example in the **mathematical register**. This needs further analysis.

**indefinite description**  An **indefinite description** is a **noun phrase** whose **determiner** is the **indefinite article** in the singular and no article or certain determiners such as **some** in the plural. It typically refers to something not known from prior discourse or the physical context.

**Example 1**  Consider this passage:

"There is a finite **group** with the property that for some proper divisor $n$ of its order a subgroup of order $n$ does not exist. However, groups also exist that have subgroups of every possible order."

" ...AND IT WAS KIND OF METAL-LIKE, YOU KNOW, WITH WHEELS, ROLLING..."

INDEFINITE DESCRIPTION.

129

The phrases "a finite group", "a subgroup", and "subgroups" are all indefinite descriptions.

An indefinite description may have a generic use, discussed under indefinite article.

**Remark 1**  This description of indefinite descriptions does not do justice to the linguistic subtleties of the concept. See [Kamp and Reyle, 1993], section 1.1.3.

**inequality**  An **inequality** is an assertion of the form $s\alpha t$, where $s$ and $t$ are terms and $\alpha$ is one of the relations $<$, $\leq$, $>$ or $\geq$. *Citations:* (24), (67), (385), (421). I have not found a citation where the relation is $\neq$ or $\ll$.

**Remark 1**  A few times, students have shown me that they were confused by this concept, since an "inequality" sounds as if it ought to mean a statement of the form $P \neq Q$, not $P < Q$ or $P \leq Q$. This is a mild case of semantic contamination.

**inert**  See mathematical object.

**infinite**  The concept of infinity causes trouble for students in various ways.

*The eternal silence of these infinite spaces fills me with dread.*
*—Blaise Pascal*

**(a) *Failure of intuition concerning size***  Students expect their intuition on size to work for infinite sets, but it fails badly. For example, a set and a proper subset can have the same cardinality, and so can a set and its Cartesian product with itself. (As Atish Bagchi pointed out to me, the intuition of experienced mathematicians on this subject failed miserably in the nineteenth century!) This is discussed further under snow.

**(b) *Infinite vs. unbounded***  Students may confuse "infinite" with "unbounded". For example, omputing science students learn about the set $A^*$ of strings of finite length of characters from an alphabet $A$. There is an *infinite number* of such strings, each one is of *finite length*, and there is *no limit* on how long they can be (except to be finite). I have seen students struggle with this complex of ideas many times.

*(c) Treating "∞" as a number.* Of course, mathematicians treat this symbol like a number in *some* respects but not in others. Thus we sometimes say that $1/\infty = 0$ and we can get away with it. Students then assume we can treat it like a number in other ways and write $\infty/\infty = 1$, which we cannot get away with. This is an example of **extrapolation**.

*References:* The mathematical concepts of infinity are discussed very perceptively in [Lakoff and Núñez, 2000], Chapter 8. Student difficulties are discussed by Tall [2001].

**infix notation** A function of two **variables** may be written with its name between the two arguments. Thus one writes $3 + 5$ rather than $+(3, 5)$. This is **infix notation**. Usually used with **binary operations** that have their own nonalphabetical **symbol**. **Relations** are written this way, too, for example "$x < y$". See **prefix notation** and **postfix notation**. *Citations:* (130), (273), (346).

**inhabit** The meaning of a **statement** in **mathematical discourse** that "*A* inhabits *B*" must be deduced from context. The **citations** show that it can mean *A* is an element of *B*, *A* is contained in the **delimiters** *B*, *A* is in the *B*th place in a list, and that *A* is a **structure** included in some sense in the space *B*. **Lives in** is used similarly in conversation, but I have found very few citations in print. See also **in**. *Citations:* (6), (16), (301), (365), (420).

*Acknowledgments:* Guo Qiang Zhang.

**injective** A function $f$ is **injective** if $f(x) \neq f(y)$ whenever $x$ and $y$ are in the domain of $f$ and $x \neq y$. Also called **one-to-one**. *Citations:* (255), (298), (392), (430).

See also **surjective**.

**Remark 1**   When proving statements using this concept, the **contra-positive** form of the definition is often more convenient.

**Remark 2**   Students often confuse this concept with the **univalent** property of **functions**: this is the property that if $(a, b)$ and $(a, b')$ are both in the graph of the function, then $b = b'$, so that the expression $f(a)$ is **well-defined**. See **in your own words**.

In the sixties there were older math-ematicians who became quite incensed if I said "injective" instead of "one-to-one". At the time I understood that this attitude was connected with an anti-Bourbaki stance. The last one that I can remember who had this attitude died re-cently. That is how language changes.

**Remark 3**   The word injective is also used in a different sense to denote a certain type of structure, as in "injective module". Searches on **JSTOR** turn up more occurrences of that usage than the one given here.

**input**   See **function**.

**insight**   You have an **insight** into some mathematical phenomenon if you have a sudden jump in your understanding of the phenomenon. This may be accompanied by ejaculations such as "**aha!**", "**eureka!**", or "**I get it!**" The jump may be in incremental (but not gradual!) increase in understanding (worthy of "aha!") or a complete leap from incomprehension to clarity ("eureka!").

**Example 1**.   The geometric diagram in the margin proves that
$$a^2 - b^2 = (a - b)(a + b)$$
at least for positive real numbers $a$ and $b$ with $b < a$.

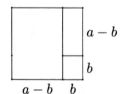

This causes some who have not seen it before to have a feeling some-thing like: "Aha! Now I really understand it!" or at least, "Aha! Now I have a better grasp of why it is true." Even if you don't feel that way about this proof, you may have experienced a similar feeling about another theorem, perhaps one whose proof by **symbol manipulation** was more obscure.

Compare this with the proof given under **symbol manipulation**. Another example of the aha experience is given under **conceptual**. In many cases, the gain in insight is irreversible, an instance of the **ratchet effect**.

**Remark 1**   It appears to me that in every case I can think of, a gain in insight as described here consists of acquiring a **metaphor** or **model** that allows one more easily to think about the problem, or visualize it. (But the metaphor need not be graphical or visual.)

**Remark 2**   In my experience, the clarity that you feel after a Eureka insight tends to become a bit cloudy as you become aware of subtleties you didn't originally notice.

*References:* [Bullock, 1994], [Frauenthal and Saaty, 1979], [Halmos, 1990], pages 574ff.

**instantiate**   To **instantiate** a free variable in an **expression** is to replace it with an identifier of a specific **mathematical object** of the appropriate **type**. If all the free variables in an expression are instantiated, the expression should denote a **specific object**.

**Example 1**   If you instantiate $x$ at 5 in the expression $2x + 1$ you get an expression denoting 11.

**integer**   A whole number, positive, negative or zero. *Citations:* (55), (78), (104), (119), (219).

**Remark 1**   I have no citation in which "integer" means specifically nonnegative integer or **positive** integer. However, students quite commonly assume that the word means nonnegative or positive integer.

See **divide**.

**integral**   This word has three different meanings.

Many computer languages are arranged so that an integer is not a real number. This may be indicated by requiring that every number be explicitly declared as one or the other, or by the **convention** that a number is real only if it is **represented** using a decimal point. For example, in Mathematica®, "32" is an integer and "32.0" is a real number. Students often assume that mathematicians follow that convention and need to be explicitly told that they don't.

### 1. Being an integer

"Integral" is used as an adjective to require that the **noun phrase** it modifies denote an integer (8 is an integral **power** of 2). *Citations:* (120), (138).

### 2. Antiderivative

An **integral** of a **function** is an **antiderivative** of the function. As such it is an operator from the set of integrable functions to the set of continuous functions **modulo** the relation of differing by a constant. In this usage the operator is often called the "indefinite integral". The indefinite integral of $f$ is denoted $\int f(x)\,dx$. One may also refer to a specific antiderivative using the form $\int_c^x f(t)\,dt$.

The word is also used to denote a solution of a more general differential equation. See also **delimiters**. *Citations:* (215), (292), (430).

### 3. Definite integral

"Integral" is also used to denote a "definite integral": this operator takes an integrable function and an **interval** (or more general **space**) on which the function is defined and produces a number. The definite integral of $f$ on an interval $[a, b]$ is denoted $\int_b^a f(x)\,dx$ or $\int_b^a f$. *Citations:* (27), (95), (138), (229).

**The integral sign as delimiter**  For many authors, the integral sign and the $dx$ in the **expression** $\int_a^b x^2 + 1\,dx$ delimit the integrand (and of course also provide other information – they are not **bare delimiters**).

However, many others do not recognize these **symbols** as delimiters and would write $\int_a^b (x^2 + 1)dx$. There is a strong **argument** for the latter position. Historically the intuition behind the expression $\int_a^b f(x)\,dx$ is that it is a *sum*: the integral sign is an elongated letter "S", and the terms of the sum are the infinitesimal rectangles with height $f(x)$ and width $dx$. This perception is the motivation for most physical applications of definite integration, so it is reasonable to teach it explicitly and to keep one's notation consistent with it.

### interpretation

An **interpretation** of a text is the current assignment of a value (possibly a **variable** object) to each **free identifier** used in the **discourse**. With a given **semantics**, the text with that interpretation may result in **assertions** about the values of the identifiers which may be true or false or (if some identifiers are **variables**) indeterminate. See **context** and **standard interpretation**.

In **mathematical logic** the language is a **formal language** and the values lie in some **mathematical structure** defined for the purpose.

### intensional

See **semantics**.

### interval

An **interval** is a subset of the set of **real numbers** of one of the following particular forms, where $a, b \in \mathbb{R}$:

- $(a, b) = \{x \mid a < x < b\}$.
- $[a, b) = \{x \mid a \leq x < b\}$.
- $(a, b] = \{x \mid a < x \leq b\}$.
- $[a, b] = \{x \mid a \leq x \leq b\}$.

**Variations** Sometimes one also uses the word "interval" for expressions of the form $(a, \infty) = \{x \mid a < x\}$ and analogous constructions. Commonly such intervals are qualified as **infinite intervals**.

One also refers to intervals of rational numbers or integers, or indeed in any partially ordered set, defined in the same way. In a non-totally-ordered set, an interval can be "fat", perhaps violating one's image of the concept. *Citations:* (67), (233), (252), (396).

### intuition

See **mental representation**.

### irregular syntax

The **symbolic language** contains some identifiers with **irregular syntax**.

***Example 1***   In the commonest usage (using **prefix notation**), the rule for applying **functions** puts the **identifier** of the function on the left of the argument and puts **parentheses** around the argumant. However, the notation "!" for the **factorial** function is put on the right and the parentheses are omitted except when necessary for grouping. *Citations:* (7), (138), (420).

***Example 2***   With some function identifiers the parentheses are conventionally omitted by most authors who otherwise use them. This causes trouble for some students learning programming languages where something like `sin(x)` or `sin[x]` is required.

*Examples:* "$\sin \pi = 0$."

"$\log \frac{3}{2} = \log 3 - \log 2$."

"$n! > 2^n$."

*Citations:* (211), (303).

***Example 3***   From the point of view of calculus students, both common notations for derivatives show irregular behavior. The **prime** notation is normally used only for functions of one **variable** but not for functions of more than one variable. Similarly, the notation involving $d$ is used for derivatives of functions of one variable; for more than one variable one must change it to $\partial$.

There are of course reasons for this. In particular, one could have in mind that $d$ is the *total* derivative operator, which coincides with the derivative when the function is of one variable; then $\partial$ is needed in the case of more than one variable because the partial derivative is a different operator. (See [Cajori, 1923], page 2.) But the beginning calculus student does not know this.

See also **orthogonal**.

***Remark 1***   These phenomena are exactly analogous to the fact that some verbs in English are irregular: for example, the past tense of "hatch" is "hatched", but the past tense of "catch" is "caught".

The irregularity of "!" lies in the **symbol**, not the meaning. If the factorial function were called Fac, then in the usual practice one would write Fac($n$), not $n$ Fac. Similarly, in English one would say I "grabbed" the ball, not using an irregular form for the past tense of "grab", even though one would use the irregular form for "catch".

Moreover, occasionally authors use the symbol "!" for some function other than the factorial (although usually analogous to it in some way), but they still write it on the right. *Citation:* (186).

**isomorphic**   Each type of **mathematical structure** has its own definition of "isomorphism". The categorists' definition of isomorphism (a morphism that has an inverse) has all these definitions as special cases. *Citations:* (398), (144).

***Difficulties***   Students frequently don't catch on to the fact that, if $M$ and $N$ are isomorphic structures of some type, there can be *many* isomorphisms between $M$ and $N$.

See **copy** and **up to**.

**italics**   A style of printing that *looks like this*. Many texts put a **definiendum** in italics. See **definition**.

**jump the fence**   If you are working with an **expression** whose **variables** are **constrained** to certain values, and you **instantiate** the expression at a value that violates the constraint, you **jump the fence**.

***Example 1***   A student, in dealing with a sum of Fibonacci numbers, might write

$$\sum_{k=0}^{n} f(k) = \sum_{k=0}^{n} f(k-1) + \sum_{k=0}^{n} f(k-2)$$

not noticing that the sums on the right involve $f(-1)$ and $f(-2)$, which may not have been defined when the definition of Fibonacci number was given.

**Terminology** The name "jump the fence" is my variation of the "fencepost error" discussed in [Raymond, 1991].

**just** One use of the word "just" in **mathematical discourse** is to indicate that what precedes satisfies the statement that occurs after the word "just".

**Example 1** (Assuming $r$ and $s$ are known to be integers greater than 1).
    " ... Then $m = rs$. But that is just the definition of "composite"."

(Or "That just means that $m$ is composite".)

**Remark 1** My own perception of this usage before I looked for citations is that the word "just" meant that what followed was *equivalent* to what preceded, but in many citations what follows is only a consequence of what precedes. Indeed, in some citations it is completely redundant.
    *Citations:* (372), (303), (340).

**just in case** This phrase means that what follows is logically **equivalent** to what precedes.

**Example 1** "An integer is even just in case it is divisible by 2."
*Citation:* (340).

**juxtaposition** Two **symbols** are **juxtaposed** if they are written down one after the other. This most commonly indicates the numerical product (but see Example 2 under **number**). Juxtaposition is also used to denote other **binary operations**, for example the operation of "and" in Boolean algebra, the concatenate of strings and the application of **trigonometric functions**. *Citations:* (94), (130).

**labeled style**   The **labeled style** of writing mathematics requires labeling essentially everything that is written according to its intent: definition, theorem, proof, remark, example, discussion, and so on. Opposed to **narrative style**.

    The most extreme examples of labeled style are **proofs**, often in geometry, that are tabular in nature with the proof steps numbered and justified by referring to previous steps by number.

    *Reference:* The labeled style was named and discussed in [Bagchi and Wells, 1998a].

**lambda notation**   A notation for referring to a **function**. The function is denoted by $\lambda x.e(x)$, where $e$ is some **expression** that allows one to calculate the value of the function at $x$. The $x$ is **bound** in the expression $\lambda x.e(x)$.

***Example 1***   "The function $\lambda x.x^2$ has exactly one critical point." This notation is used in **mathematical logic**, computing science, and linguistics, but not generally by mathematicians. *Citation:* (45).

    Compare **barred arrow notation**.

**large**   A text that says one set is **larger** than another may be referring to the inclusion ordering, or may be referring to **cardinality**. *Citations:* (373), (409). Note that in the second citation the authors feel obligated to explain that they mean cardinality, not inclusion.

    "Large" said of a number can mean large positive or large negative, in other words large in absolute value. *Citations:* (51), (300).

**law of gravity for functions**   In mathematics at the graduate level the student may notice that **functions** are very often illustrated or visualized as **mapping** the **domain** *down* onto the **codomain**. Any function in fact produces a structure on the domain (the quotient space) that in the case of some kinds of structures (sheaves, Riemann surfaces, and others) is quite elaborate. In those applications the function is often thought of as a projection. This is presumably the motivation for the use of the word **under**.

In teaching such courses I have found it helpful to point out this phenomenon to students, who from calculus may visualize functions as going up, and from discrete mathematics or abstract algebra may think of them as going from left to right. See **over** and **rightists**.

**lemma**  A theorem. One may typically expect that a lemma is not of interest for itself, but is useful in proving other theorems. However, some lemmas (König's Lemma, Schanuel's Lemma, Zorn's Lemma) have become quite famous.

*Acknowledgments:* Owen Thomas. *Citations:* (46), (122), (260), (345).

**lemmata**  Lemmas. An obsolete plural.

**let**  "Let" is used in several different ways in the **mathematical register**. What follows is a tentative classification. Some of the variations in usage (as in Examples 1 and 2) make no difference to the logical **argument** that the usage expresses. This may make the classification seem excessively picky. I am not aware of research on students' misunderstandings in these situations.

In many cases, **assume**, **suppose** and if can be used instead of "let". The syntax for these others is different; thus one says "Let $x$ *be* ... " but "Assume [Suppose] $x$ *is* ... " Also, "If $x = 1$" cannot be a complete sentence, but "Let $x = 1$" can be. The words "assume" and "if" are used in some situations where "let" is inappropriate; those usages are discussed under **assumption** and if. There are other subtle differences about the way "assume", "suppose", "let" and "if" are used that need further investigation.

*(a) Introducing a new symbol or name* One common use of "let" is to introduce a new **symbol** or **name**. This, of course, is a species of **definition**, usually with a restricted **scope** (the current section of the text, for example).

**Example 1** Consider the theorem
"An integer divisible by 4 is divisible by 2."
A proof could begin this way:
"Let $n$ be an integer divisible by 4."
This introduces a new **variable** symbol $n$ and **constrains** it to be divisible by 4.

**Example 2** Suppose the theorem of the preceding example had been stated this way:
*"Let $n$ be an integer. If $n$ is divisible by* 4 *then it is divisible by* 2."
Then the proof could begin
"Let $n$ be divisible by 4."
In this sentence, $n$ is introduced in the theorem and is further constrained in the proof.

**Remark 1** These two examples illustrate that whether a new symbol is introduced or a previous **symbol** is given a new **interpretation** is a minor matter of wording; the underlying logical structure of the **argument** is the same.

**Remark 2** "Define" is sometimes used in this sense of "let"; see Example 2 under **mathematical definition**. Of course, there is no *logical* distinction between this use of "let" and a formal definition; the difference apparently concerns whether the newly introduced expression is for temporary use or **global** and whether it is regarded as important or not. Further investigation is needed to spell the distinction out.

If, assume and suppose can be used in this situation, with requisite changes in syntax: "is" instead of "be" for assume and suppose, and the sentences must be combined into one sentence with "if". *Citations:* (46), (155), (69).

**(b)** *To consider successive cases*

**Example 3** "Let $n > 0$. ... Now let $n \leq 0$. ... " If, assume and suppose seem to be more common that "let" in this use. See now. *Citations:* (12), (109).

**(c)** *To introduce the* precondition *of a* definition

**Example 4** "**Definition** Let $n$ be an integer. Then $n$ is **even** if $n$ is divisible by 2." If, assume and suppose can be used here. *Citations:* (56), (128), (325).

**(d)** *To introduce an arbitrary object* To pick an unrestricted object from a collection with the purpose of proving an assertion about all elements in the collection using universal generalization. Example 1 above is an example of this use. Often used with arbitrary. If, assume and suppose can be used here. *Citations:* (143), (53)

**(e)** *To name a witness* To provide a local identifier for an arbitrary object from a collection of objects known to be nonempty. Equivalently, to choose a witness to an existential assertion that is known to be true. If, assume and suppose can be used here.

**Example 5** In proving a theorem about a differentiable function that is increasing on some interval and decreasing on some other interval, one might write:

"Let $a$ and $b$ be real numbers for which $f'(a) > 0$ and $f'(b) < 0$."
These numbers exist by hypothesis.

**Example 6**   In the context that $G$ is known to be a noncommutative group:

> "Let $x$ and $y$ be elements of $G$ for which $xy \neq yx \ldots$"

The following is a more explicit version of the same assertion:

> "Let the noncommutative group $G$ be given. Since $G$ is noncommutative, the collection $\{(x, y) \in G \times G \mid xy \neq yx\}$ is nonempty. Hence we may choose a member $(x, y)$ of this set $\ldots$ "

**Example 7**   In proving a function $F : S \to T$ is injective, one may begin with "Let $x, x' \in S$ be elements for which $F(x) = F(x')$". These elements must exist if $F$ is non-injective: in other words, this begins a proof by contrapositive. The existential assertion which the elements $x, x' \in S$ are witnesses is implied by the assumption that $F$ is not injective.

**Remark 3**   The choice of witness may be a *parametrized choice*: Given that $(\forall x)(\exists y)Q(x, y)$ and given $c$, let $d$ be an object such that $Q(c, d)$.

**Example 8**   Assuming $c$ is a complex number:

> "Let $d$ be an $n$th root of $c$."

*Citations:* (43), (176), (257), (298), (430).

**(f) "Let" in definitions**   Let can be used in the defining phrase of a definition.

**Example 9**   "Let an integer be even if it is divisible by 2."

**Remark 4**   This usage strikes me as unidiomatic. It sounds like a translation of a French ("Soit $\ldots$ ") or German ("Sei $\ldots$ ") subjunctive. If, assume and suppose cannot be used here.

*References:* This entry follows the discussion in [Bagchi and Wells, 1998a]. See also [Selden and Selden, 1999].

*Acknowledgments:* Atish Bagchi, Owen Thomas.

**lg**   See logarithm.

**life isn't fair**   My students (and I as well) sometimes feel that certain situations in mathematics *just aren't fair.*

**Example 1**   While it is true that if $W$ is a sub-vector space of $V$ then any basis of $W$ is contained in some basis of $V$, it is *not* true that any basis of $V$ contains a basis of $W$. (See **false symmetry**.)

**Example 2**   If you run around the unit circle in the complex plane evaluating the square root **function**, when you get back to where you started you have a different **value** from the one you had at the start.

**Example 3**   If you have a group $G$ with normal subgroup $K$ and a group $G'$ with a normal subgroup $K'$, and if the quotients $G/K$ and $G'/K'$ are isomorphic, it could still happen that $G$ and $G'$ are not isomorphic.

**Example 4**   No matter how hard I try, I can't find a formula in terms of functions I already know for the **integral** of $e^{x^2}$.

**Example 5**   There are *entirely too many* different kinds of function spaces.

The teacher can pontificate about how all these unfair situations give rise to *interesting mathematics*, but perhaps this should not be done right when students have lost points on a quiz because they didn't understand such booby traps.

**limit**   Students have numerous problems associated with limits.

**Example 1**   Many students believe the **myth** that a sequence that has a limit "approaches the limit but never get there". They have presumably **constructed** their **prototypical** sequence based on the examples they have seen in class or in the text, most of which behave that way. This is a form of **cognitive dissonance**. The two forms of learning required by the definition of cognitive dissonance are the definition versus the normal way we learn concepts via **generalization** from examples. *Reference:* This example is from the discussion in [Tall, 1992b], Sections 1.5 and 1.6.

***Example 2***   The $\epsilon$-$\delta$ definition of limit is complicated and difficult for students to grasp. The difficulties are similar to those discussed under **continuity**.

See [Pimm, 1983], [Cornu, 1992], [Tall and Vinner, 1981], [Tall, 1993].

**literalist**   A **literalist** or **fundamentalist** believes that the formalism used to give a **mathematical definition** or to axiomatize a set of mathematical phenomena should be taken as the "real meaning" of the idea and in extreme cases even as the primary way one should think about the concepts involved.

***Example 1***   In the study of the foundations of mathematics, one of the problems is to show that mathematics is consistent. One standard way to do this is to define everything in terms of **sets** (so that math is consistent if set theory is consistent). In particular, a **function** is defined as a set of ordered pairs, an ordered pair $(a, b)$ is defined to be $\{a, \{a, b\}\}$, and the nonnegative **integers** may be defined recursively:

$$0 = \{\} \quad \text{the } \mathbf{empty\ set}$$
$$1 = \{0\} = \{\{\}\}$$
$$2 = \{0, 1\} = \{\{\}, \{\{\}\}\}$$
$$3 = \{0, 1, 2\} = \{\{\}, \{\{\}\}, \{\{\}, \{\{\}\}\}\}$$

and so on.

Many mathematicians (but not all) would agree that it is desirable to do this *for the purposes of foundations*. (see **reductionist**). A **literalist** will insist that this means that an ordered pair and the number 3 *really are* the sets just described, thus turning a perfectly legitimate consistency **proof** into a pointless statement about reality.

That sort of behavior is not damaging as long as one does not engage in it in front of students (except in a foundations class). Is it a good idea to send students out in the world who believe the **assertion** "2 is even" is

*There is no surer way to misread any document than to read it literally.*
                    *–Learned Hand*

on a par with the assertion "2 is an element of 3 but not an element of 1"?
*Reference:* A good place to read about defining integers in terms of sets is [Simpson, 2000].

***Example 2*** Literalists object to regarding an **equivalence relation** and its associated **partition** as the "same **structure**". They say things like "How can a set of subsets be the same thing as a relation?" It seems to me that this literalist **attitude** is an *obstruction to understanding the concept.*

The mature mathematician thinks of a type of structure as a whole rather than always coming back to one of the defining aspects. Students don't always get to that stage quickly. The set of subsets and the relation are merely *data* used to describe the structure. To understand the structure properly requires understanding the important **objects** and concepts (such as a function being compatible with the partition) involved in these structures and all the important things that are true of them, on an equal footing (as in the concept of clone in universal algebra), and the ability to focus on one or another aspect as needed.

***Example 3*** First order **logic** is a mathematical model of **mathematical reasoning**. The literalist attitude would say: Then the expression of our mathematical reasoning should look like first order logic. This is **esilism**.

One could argue that "fundamentalist" should mean being literal-minded about foundational definitions, but not necessarily about other definitions in mathematics, and that "literalist" should be used for the more general meaning.

***Example 4*** Literalists may also object to phrases such as "incomplete proof" and "this **function** is not **well-defined**". See **radial concept**.

See also the discussions under **mathematical definition**, **mathematical logic** and **mathematical structure**.

*Acknowledgments:* Eric Schedler, Peter Freyd, Owen Thomas and also [Lewis and Papadimitriou, 1998], page 9, where I got the word "fundamentalist". Lewis and Papadimitriou did not use the word in such an overtly negative way as I have.

*References:* [Benaceraff, 1965], [Lakoff and Núñez, 1997], pages 369–374, and [Makkai, 1999].

**lives in**   See inhabit.

**ln**   See logarithm.

**local**   With respect to a **structure** $\mathfrak{M}$, an **object** is defined **locally** if it is in some sense defined only on a substructure of $\mathfrak{M}$. It is defined **globally** if it is defined on all of $\mathfrak{M}$. This usage is usually informal, but in some cases the word "local" or "global" has a formal definition. *Citations:* (58), (65), (84), (170), (205), (418).

The words may be used in settings outside the **mathematical register**. For example, one might complain that one understands a **proof** "locally but not globally", meaning that one can follow the individual steps but has no overall grasp of the proof.

***Example 1***   The phrases **local identifier** and **global identifier** in this text (borrowed from computing science) are examples of informal usage of the terms.

*Citations:* (60).

**local identifier**   A local identifier in a segment of a mathematical **text** is an **identifier** for a particular **mathematical object** that has that meaning only in the current block of text. The block of text for which that meaning is valid is called the **scope** of the identifier.

The scope may be only for the paragraph or subsection in which it is defined, with no explicit specification of the scope given. One clue for the reader is that **definitions** with this sort of restricted scope are typically introduced with words such as **let** or **assume** rather than being given the formal status of a definition, which the reader tends to assume will apply to the rest of the **discourse**.

The author may make the scope explicit.

***Example 1***   "Throughout this chapter $f$ will be a continuous function." ***Citations:*** (178), (213), (271), (285).

See also global identifier.

## location

***(a) Physical location***   In mathematical discourse, words such as where, **anywhere** and **wherever**, and local prepositions such as "in" and so on are used to refer to physical locations in the same way that they are used in ordinary discourse. By extension, they are used to refer to locations in a particular discourse.

***Example 1***   "In this section, $\phi$ is a continuous function." (See also local identifier.) This is a normal part of academic discourse.

***(b) Location in a structure***   In mathematics, it is common to refer to a subset of elements of any set as if it were a location.

***Example 2***   "The function $\phi$ is **positive** wherever its derivative is positive."

This presumably originated from the many examples where the set in question is the set of points of a **space** (see **mathematical structure**.)

See also **time**. ***Citations:*** (153), (254), (312).

**logarithm**   The expression "$\log x$" has a **suppressed parameter**, namely the base being used. My observation is that in pure mathematics the base is normally $e$, in texts by scientists it may be 10, and in computing science it may be 2, and that in all these cases the base may not be explicitly identified.

Students in particular need to know that this means there are *three different functions* in common use called "log". See also **trigonometric functions**.

***Remark 1***   In calculus texts, $\log_e$ may be written "ln", and in computing science $\log_2$ may be written "lg". ***Citations:*** (25), (169), (264).

*Acknowledgments:* Owen Thomas.

**look ahead**   When performing a calculation to solve a problem, one may **look ahead** to the form the solution must take to guide the manipulations one carries out.

***Example 1***   Given a right triangle with legs $a$ and $b$ and hypoteneuse $c$, one can derive the Pythagorean Theorem $a^2 + b^2 = c^2$ from the identity

$$\sin^2 \theta + \cos^2 \theta = 1$$

by rewriting it as

$$\frac{a^2}{c^2} + \frac{b^2}{c^2} = 1$$

and then multiplying by $c^2$. Olson [1998] discovered that when asked to reverse the process to derive the **trig identity** from the Pythagorean Theorem, some students balked at the first step, which is to divide the equation $a^2 + b^2 = c^2$ by $c^2$, because "there is no reason to divide by $c^2$": The students apparently could think of no method or **algorithm** which said to do this. Of course there *is* a method — **look ahead** to see what form of the equation you need. More about this example in section (a) under **attitudes**. This is related to **walking blindfolded**.

**lowercase**   See **case**.

**Luddism**   **Luddism** is an unreasoning opposition to all technological innovation. Luddites appear in mathematics, most noticeable lately concerning the use of calculators and computers by students. There is also resistance to new terminology or notation.

***Example 1***   There is a legitimate debate over such questions as: Should calculators be withheld from students until they can do long division rapidly and accurately? Should Mathematica be withheld from students until they can carry out formal integration rapidly and accurately? Unfortunately, professors by their nature tend to be skilled in argumentation, so it may take long anthropological observation to distinguish a Luddite from a rational opponent of a particular piece of technology.

***Remark 1*** The two questions in the preceding remark do not have to be answered the same way. Nor do they have to be answered the same way for math majors and for other students.

**macron** See bar.

**malrule** A **malrule** is an incorrect rule for **syntactic** transformation of a mathematical **expression**. Examples are given in the entry for **extrapolate**.

This name comes from the mathematics education literature.

**map** Also **mapping**. Some texts use it interchangeably with the word "function". Others distinguish between the two, for example requiring that a mapping be a **continuous** function. See **function**. *Citations:* (200), (252), (298), (357).

*Acknowledgments:* Michael Barr.

**marking** See definition.

**matchfix notation** See outfix notation.

**mathematical education** One purpose of this Handbook is to raise mathematicians' awareness of what specialists in mathematical education have found out in recent years. The following entries discuss that and have pointers to the literature.

abstraction 7
and 15
APOS 17
attitudes 22

behaviors 25
cognitive dissonance 36
compartmentalization 39
concept 41

# mathematical education

# mathematical logic

The Handbook's website provides links to some resources in mathematical education.

**mathematical logic**  **Mathematical logic** is any one of a number of **mathematical structures** that **models** many of the **assertions** spoken and written in the **mathematical register**. Such a structure typically is provided with rules for **proof** and rules for giving meaning to items in the structure (**semantics**). The phrases **formal logic** and **symbolic logic** are also used.

***First Order Logic***  The most familiar form of mathematical logic is **first order logic**, in which, as in many other forms of logic, sentences are represented as strings of symbols. For example,

"There is an $m$ such that for all $n$, $n < m$"

would typically be represented as "$\exists m \forall n \, (n < m)$".

*Logic is the hygiene the
mathematician prac-
tices to keep his ideas
healthy and strong.*
        *–Hermann Weyl*

First order logic is a useful codification of many aspects of mathematical formalism, but it is not the only possible result of any attempt of formalizing mathematics. The website [Zal, 2003] lists many types of logic, some stronger than first order logic and some weaker, designed for use with mathematics, computing science, and real-world applications. The approach of category theory to model theory, as expounded in [Makkai and Reyes, 1977], [Makkai and Paré, 1990], and [Adámek and Rosický, 1994], produces formal systems that are very different in character from standard first order logic and that vary in strength in both directions from first order logic. However, first order logic has many nice formal properties and seems particularly well adapted to mathematics.

Some mathematicians operate in the belief that the assertions and **proofs** they give in the **mathematical register** can in principle be **translated** into first order logic. This is desirable because in theory a purported proof in the formal symbolism of mathematical logic can be mechanically checked for correctness. The best place to see the **argument** that every mathematical proof can *in principle* be translated into first order logic is the book [Ebbinghaus, Flum and Thomas, 1984] (read the beginning of Chapter XI). In particular, proofs involving quantification over sets can be expressed in first order logic by incorporating some set of axioms for set theory.

In practice no substantial proof gets so far as to be expressed in logical symbolism; in fact to do so would probably be impossibly time-consuming and the resulting proof not mechanically checkable because it would be too large. What does happen is that someone will challenge a step in a proof and the author will defend it by expanding the step into a proof containing more detail, and this process continues until everyone is satisfied. The mathematicians mentioned in the preceding paragraph may believe that if this expansion process is continued long enough the proof will become a proof in the sense of mathematical logic, at least in

152

the sense that every step is directly translatable into logical formalism. Even if this is so, caveats must be attached:

- First order logic may be optimal for **mathematical reasoning**, but not for reasoning in everyday life or in other sciences.
- First order logic is clearly not the ideal language for *communicating* mathematical arguments, which are most efficiently and most clearly communicated in the **mathematical register** using a mixture of English and the **symbolic language**.

***Geometric and other insights***   Aside from those caveats there is a more controversial point.  Consider the **proof** involving the monk given in Example 2 under **conceptual blend**. This proof can probably be transformed into a proof in first order logic (making use of continuous mappings and the intermediate value theorem), but the resulting proof *would not be the same proof* in some sense. In particular, it loses its physical immediacy. Many geometric proofs as well have a (physical? visual?) immediacy that is lost when they are translated into first order logic.

One could defend the proposition that all proofs can be translated into first order logic by either denying that the monk proof (and a pictorial geometric proof) is a mathematical proof, or by denying that the translation into first order logic changes the proof. The first approach says many mathematicians who think they are doing mathematics are not in fact doing so. The second violates my own understanding of how one does mathematics, because what is lost in the translation is for me *the heart of the proof.* Specifically, the checking one could do on the first order logic form of the proof would not check the physical or geometric content.

Nevertheless, the translation process may indeed correctly **model** one sort of proof as another sort of proof. It is a Good Thing when this can be done, as it usually is when one kind of mathematics is modeled in another. My point is that the two kinds of proof are different and both must be regarded as mathematics. See the discussion in [Tall, 2002].

There is more about the suitability of mathematical logic in Remark 2 under esilism.

See order of quantifiers, translation problem, and esilism.

*References:* First order logic is presented in the textbooks [Mendelson, 1987], [Ebbinghaus, Flum and Thomas, 1984], [van Dalen, 1989]. The formalisms in these books are different but equivalent. The book [Lakoff and Núñez, 2000], Chapter 6, discusses logic from the point of view of cognitive science.

*Acknowledgments:* Discussions with Colin McLarty.

**mathematical mind**   People who have tried higher level mathematical courses and have become discouraged may say, "I just don't have a mathematical mind" or "I am **bad at math**". Some possible reasons for this attitude are discussed under ratchet effect, trivial and yes it's weird. Reasons for people being discouraged about mathematics (or hating it) are discussed in [Kenschaft, 1997]. See also [Epp, 1997].

I do not deny that some people have a special talent for mathematics. In particular, the best undergraduate mathematics students tend not to have most of the difficulties many students have with abstraction, proof, the language used to communicate mathematical reasoning, and other topics that take up a lot of space in this book. It appears to me that:

- We who teach post-calculus mathematics could do a much better job explaining what is involved in abstract mathematics, to the point where many more students could get through the typical undergraduate abstract algebra or analysis course than do now.
- The students who need a lot of such help are very likely not capable of going on to do research at the Ph.D. level in mathematics. If this is correct, the people who believe in **osmosis** would be correct –

if the only students who studied mathematics were future research mathematicians!

These two points do not contradict each other. They are both factual claims that could be tested by a (long, expensive) longitudinal study.

Platonism 197
time 251

**mathematical object**  Mathematical objects are what we refer to when we do mathematics. *Citation:* (121),

**(a) *The nature of mathematical objects***  Mathematicians talk about mathematical objects using most of the same grammatical constructions in English that they use when talking about physical objects (see **Platonism**). In this discussion I will take *the way we talk about mathematical objects* as a starting point for describing them. I will not try to say *what they really are* but rather will observe some properties that they must have given the way we talk about them.

*(i) Repeatable Experience*  Mathematical objects are like physical objects in that our experience with them is *repeatable*: If you ask some mathematicians about a property of some particular mathematical object that is not too hard to verify, they will generally agree on what they say about it, and when there is disagreement they commonly discover that someone has made a mistake or has misunderstood the problem.

*Reality is nothing but a collective hunch.*
*–Lily Tomlin*

*(ii) Inert*  Mathematical objects are **inert**. They do not change over **time**, and they don't interact with other objects, even other mathematical objects. Of course, a particular function, such as for example $s(t) = 3t^2$, may *model* a change over time in a physical object, but the function itself is the same every time we think of it.

*(iii) Eternal*  Mathematical objects are **eternal**. They do not come into and go out of existence, although our *knowledge* of them may come and go.

**Example 1** A dentist may tell you that he has a hole in his schedule at 3PM next Monday; would you like to come then? That hole in his schedule is certainly not a physical object. It is an **abstract object**. But it is not a mathematical object; it interacts with physical objects (people!) and it changes over time.

**Example 2** A variable, say `Height`, in a computer program is an abstract object, but it is not a mathematical object. At different times when the program is running, it may have different values, so it is not inert. It may be in a subroutine, in which case it may not exist except when the subroutine is running, so it is not eternal. And it can certainly interact (in a sense that would not be easy to explicate) with physical objects, for example if it keeps track of the height of a missile which is programmed to explode if its height becomes less than 100 meters.

**(b) Types of mathematical objects** It is useful to distinguish between **specific** mathematical objects and **variable** ones.

**Example 3** The number 3 is a specific mathematical object. So is the sine function (once you decide whether you are using radians or degrees). But this is subject to disagreement; see **unique**.

**Example 4** If you are going to prove a theorem about functions, you might begin, "Let $f$ be a continuous function", and in the proof refer to $f$ and various objects connected to $f$. From one point of view, this makes $f$ a **variable mathematical object**. (A logician would refer to the **symbol** $f$ as a variable, but mathematicians in general would not use the word.) This is discussed further under **variable**.

**(c) Difficulties** A central difficulty for students beginning the study of mathematics is being able to conceive of objects such as the sine function *as an object*, thus reaching the third stage of the **APOS** theory. This is the problem of **encapsulation**. Students also confuse a **mathematical object** with the **symbols** denoting it. [Pimm, 1987] discusses this in children,

156

pages 17ff, and much of the mathematical education literature concerning function mentions that problem, too, as well as the more severe problem of encapsulation. See also Example 2 under literalist.

Some of the difficulties students have when reasoning about mathematical objects may have to do with the properties we regard them as having. The difficulties students have with conditional sentences may be related to the inert and eternal nature of mathematical objects discussed previously in this entry; this is discussed further under only if and contrapositive.

*Acknowledgments:* I learned the idea that mathematical objects are inert and eternal from [Azzouni, 1994]. The example of the hole in the schedule comes from [Hersh, 1997b], page 73. Michael Barr made insightful comments. See also [Sfard, 2000a] and [Sfard, 2000b].

**mathematical register**   This is a special register of the English language used for mathematical exposition: communicating mathematical definitions, theorems, proofs and examples. Distinctive features of the mathematical register of English include

a) Ordinary words used in a technical sense, for example, "function", "include", "integral", and "group".

b) Technical words special to the subject, such as "topology", "polynomial", and "homeomorphism".

c) Syntactic structures used to communicate the logic of an argument that are similar to those in ordinary English but with *differences in meaning*. This list describes those structures discussed in this book:

*Mathematicians are like Frenchmen: whatever you say to them they translate into their own language and forthwith it is something entirely different. – Johann Wolfgang von Goethe*

Any register belonging to a technical subject has items such as (a) and (b). Some words like these are listed in this Handbook, including words that cause special problems to students and words that are used with **multiple meanings**.

The syntactic structures mentioned in (c) are a major stumbling block for students. It appears to me that these structures make the mathematical register quite unusual even among technical registers in general in how far its **semantics** deviates from the semantics of ordinary English. (However, every tribe thinks it is "more different" than any other tribe ... ). Some of these syntactic structures involve English expressions that are used with meanings that are subtly different from their meanings in ordinary English or even in the general scientific register.

*References:* There seem to be very few articles that study the mathematical register specifically. A brief overview is given by [Ferrari, 2002]. Some aspects are described in [Epp, 1999], [Pimm, 1987], [Pimm, 1988], [Schweiger, 1994a], [Schweiger, 1994b], [Schweiger, 1996]. Steenrod [1975], page 1, distinguishes between the mathematical register (which he calls the "formal structure") and other registers.

N. J. de Bruijn [1994] introduces the concept of the **mathematical vernacular**. He says it is "the very precise mixture of words and formulas used by mathematicians in their better moments". He excludes some things, for example **proof by instruction** ([de Bruijn, 1994], page 267), which I would include in the mathematical register. He makes a proposal for turning a part of the mathematical vernacular into a formal system and in the process provides a detailed study of part of (what I call the)

*Mathematics is written for mathematicians.*
*–Nicolaus Copernicus*

mathematical register as well as other types of mathematical writing.

Many mathematical texts include discussions of history, intuitive descriptions of phenomena and applications, and so on, that are in a general scientific register rather than the mathematical register. Some attempts to classify such other types of mathematical writing may be found in [Bagchi and Wells, 1998a], [de Bruijn, 1994], and in Steenrod's article in [Steenrod *et al.*, 1975].

*Reference:* Much of the current discussion is drawn from [Bagchi and Wells, 1998a].

*Acknowledgments:* Cathy Kessel.

**mathematical structure**   A **mathematical structure** is a **set** (or sometimes several sets) with various associated **mathematical objects** such as subsets, sets of subsets, operations of various **arities**, and **relations**, all of which must satisfy various requirements. The collection of associated **mathematical objects** is called the **structure** and the set is called the **underlying set**.

> The definition given here of mathematical structure is not a **mathematical definition**. To give a proper mathematical definition of "mathematical structure" as a set with structure results in an unintuitive and complicated construction.

Two examples of definitions of mathematical structures may be found under **equivalence relation**. The examples given there show that the same structure can have two very different definitions.

***Example 1***   A topological **space** is a set $S$ together with a set $T$ of subsets of $S$ satisfying certain requirements.

Presenting a complex mathematical idea as a mathematical structure involves finding a minimal set of associated objects (the structure) and a minimal set of conditions on those objects from which the theorems about the structure follow. The minimal set of objects and conditions may not be the most important aspects of the structure for applications or for one's **mental representation**. See **definition**.

***Example 2***   A function is commonly defined as a set of ordered pairs with a certain property. A mathematician's picture of a function has many facets: how it models some covariation (for example, velocity), its behavior in the limit, **algorithms** for calculating it, and so on. The set of ordered pairs is not what first comes to mind, except perhaps when one is thinking of the function's graph.

The word "structure", sometimes in the phrase "mathematical structure", is also used to describe the way certain types of **mathematical objects** are related to each other in a system. This sense is similar to the meaning of **schema**.

***Example 3***   One could investigate the structure of the solutions of a particular type of differential equation.

*Citations:* (179), (364), (368), (406). See also **space**.

**maximize**   To **maximize** a function is to find **values** of its **argument** for which the function has a maximum. **Minimize** is used similarly.

The **metaphor** behind this usage seems to be: vary the **input** over **time** until you find the largest value. *Citations:* (135), (247), (316).

**mean**

### 1. To form a definition

"Mean" may be used in forming a **definition**. *Citations:* (3), (173).

***Example 1***   "To say that an integer is **even** means that it is divisible by 2."

### 2. Implies

To say that an **assertion** $P$ means an assertion $Q$ may mean that $P$ **implies** $Q$. *Citation:* (372).

***Example 2***   "We have proved that 4 divides $n$. This means **in particular** that $n$ is even."

**Remark 1**  Of course, **mean** is also a technical term (the average).

**member**  See element.

**mental representation**  One's **mental representation** (also "internal representation") of a particular mathematical concept is the cognitive structure associated with the concept, built up of **metaphors**, mental pictures, **examples**, properties and processes related to each other by **conceptual blending** and in other ways.

The mental representation is called the **concept image** by many writers in **mathematical education**. The definition just given is in fact a modification of the definition of concept image given by Tall [1992b], page 7. The way human concepts are organized, as described by cognitive scientists such as George Lakoff [Lakoff, 1986], includes much of the structure of the mental representation of the concept in my sense. This is discussed further under **concept**, **prototype**, **radial concept** and **schema**.

In written or spoken **mathematical discourse**, discussion of some aspect of the mental representation of a concept is often signaled by such phrases as **intuitively** or "you can think of ... ". *Citation:* (181).

*(a) Mental representations and definitions*  The contrast between a student's mental representation of a concept and its mathematical **definition** is a source of **cognitive dissonance**. Students may avoid the disparity by ignoring the definition. The disparity comes about from inappropriate learning strategies such as **generalization** and **extrapolation**.

> I have known both logicians and computer scientists (but not many, and no mathematicians) who deny having any nonsymbolic mental representations of mathematical concepts. Some of them have claimed to be entirely syntax directed; all they think of is **symbols**. Perhaps some of these colleagues do have mental representations in the broad sense, but not pictorial or geometric ones. Possibly the phrase "mental image" should be restricted to cases where there is geometric content.

Professional mathematicians who are learning a subject know they must *adjust their mental representation to the definition*. In contrast, in doing research they often quite correctly *adjust the defini-*

*Half this game
is 90% mental.*

*–Yogi Berra*

*The folly of mistaking
a paradox for a dis-
covery, a metaphor for
a proof, a torrent of
verbiage for a spring
of capital truths, and
oneself for an ora-
cle, is inborn in us.*

*–Paul Valery*

*tion* instead of their mental representation. That is a primary theme of [Lakatos, 1976].

*References:* Many articles in the book [Tall, 1992a] discuss mental representation (under various names often including the word "image") in depth, particularly [Tall, 1992b], [Dreyfus, 1992] and [Harel and Kaput, 1992]. See also [Dieudonné, 1992] V.6, page 163, [Kieran, 1990], [Meel, 1998] (especially pages 168–170), [Piere and Kieren, 1989], [Presmeg, 1997a], [Thompson and Sfard, 1998], [Tall and Vinner, 1981], [Wells, 1995], [Wheatley, 1997].

Mental imagery is discussed from a philosophical point of view, with many references to the literature, by Dennett [1991], Chapter 10. The book [Lakoff, 1986] is concerned with concepts in general, with more of a linguistic emphasis.

A sophisticated mental representation of an important concept will have various formalisms and mental pictures that fit together by **conceptual blending** or **metaphor**. [Lakoff and Núñez, 1997] regard metaphor as central to understanding what mathematics is all about.

See also **aha**, **conceptual**, **mathematical object**, **Platonism** and **representation**.

**metaphor**    A **metaphor** is an implicit conceptual identification of part of one type of situation with part of another. The word is used here to describe a type of *thought configuration*, a form of **conceptual blend**. The word is also used in rhetoric as the name of a *type of figure of speech* – a linguistic entity which of course corresponds to a conceptual metaphor. (Other figures of speech, such as simile and **synecdoche**, correspond to conceptual metaphors as well.)

Lakoff and Núñez [1997], [1998], [2000] divide metaphors in mathematics into two fundamental types: **grounding metaphors**, based on everyday experience, and **linking metaphors** that link one branch of mathematics to another.

***Example 1***   The interior of a closed curve or a sphere is called that because it is like the interior in the everyday sense of a bucket or a house. This is a grounding metaphor. It also illustrates the fact that **names** in mathematics are often based on metaphors. The fact that the boundary of a real-life container has thickness, in contrast to a closed curve or a sphere, illustrates my description of a metaphor as identifying *part* of one situation with part of another. One aspect is emphasized; another aspect, where they may differ, is ignored.

***Example 2***   The representation of a number as a location on a line, and more generally tuples of numbers as locations in a **space**, *links* numbers to geometry. The mathematical concept of line is *grounded* in the everyday notion of path.

***Example 3***   The insight in the previous example got turned around in the late nineteenth century to create the metaphor of **space** as a set of points. Topology, differential geometry, and other branches of mathematics were invented to turn this metaphor into a **mathematical definition** that made the study of spaces more rigorous but also less intuitive. This is discussed further under **space**. See also **conceptual blend**, **snow** and the sidebar under **mathematical structure**.

***Example 4***   The name "set" is said to be grounded in the metaphor of "set as container". This has at least two problems caused by the fact that some aspects of real world containers don't carry over to the mathematics. For one thing, the intersection of two sets $A$ and $B$ is defined to be the set that contains those elements that are in both $A$ and $B$. Thus an element of $A \cap B$ is also in $A$ and in $B$. In real life, the elements *are not in three different containers*. Another discrepancy is that, for example, your wallet may contain a credit card, and the wallet may be contained in your pocket. In normal conversation, you would say the credit card is in your pocket. In mathematics, however, $\epsilon$ is not transitive. I have seen both these gaps cause trouble for students.

***Remark 1***    The concept of set may indeed have been historically grounded in the concept of container. However, that metaphor has never played much part in my thinking. For me, the set is a **mathematical object** distinct from its elements but completely determined by them. *It is the unique such object of type "set"*. I visualize the set as a node connected by a special relationship to exactly each of its elements and nothing else.

The description of a set as for example $\{1, 2, 3\}$ is of course also determined by the elements of the set, but the description is a linguistic object, not a mathematical one. Of course, in **mathematical logic**, linguistic objects are **modeled** by mathematical objects; in particular, one could have a **term** in a logical theory that corresponds to "$\{1, 2, 3\}$" whose **denotation** is the set. But the linguistic object, the term in the theory, and the set denoted by $\{1, 2, 3\}$ are three different things. This is not always made clear in logic texts.

***Example 5***    In **college** level mathematics we have another metaphor: set as **object** which can be the subject of **operations**. This is a linking metaphor (set as element of an algebra). This causes difficulties for students, particularly "set as element of a set"; see **object-process duality**. This is discussed in [Hazzan, 1999].

***Example 6***    One metaphor for the real line is that it is a set of points (as in Example 3). It is natural to think of points as tiny dots; that is the way we use the word outside mathematics. This makes it natural to think that to the left and right of each point there is another one, and to go on and wonder whether two such neighboring points touch each other. It is valuable to think of the real line as a set of points, but the properties of a "line of points" just described *must be ignored* when thinking of the real line. In the real line there is no point next to a given one, and the question of two points touching brings inappropriate physical considerations into an abstract structure.

This example comes from [Lakoff and Núñez, 2000]. See also **space**.

*Difficulties* Most important mathematical concepts are based on several metaphors, some grounding and some linking; for examples see the discussion under **function**. These metaphors make up what is arguably the *most important part* of the mathematician's mental representation of the concept. The daily use of these metaphors by mathematicians cause enormous trouble to students, because each metaphor provides a way of thinking about an *A* as a kind of *B in some respects*. The student naturally thinks about *A* as a kind of *B* in inappropriate respects as well.

Students also notoriously have difficulty in switching between different suitable metaphors for the same **object**; possessing that ability is a reliable sign of a successful student of mathematics.

The discussion in Example 6 is the tip of an iceberg. It may be that *most* difficulties students have, especially with higher-level mathematics (past calculus) are based on *not knowing which aspects of a given metaphor are applicable in a given situation,* indeed, on not being consciously aware that one has to restrict the applicability of the mental pictures that come with a metaphor.

Why not tell them? It would be appropriate for textbooks to devote considerable space to how mathematicians think of each concept, complete with a discussion of which aspects of a metaphor are apt and which are not.

*References:* See [Lanham, 1991] for figures of speech and [Lakoff and Núñez, 2000], Chapter 2, for an introduction to metaphors in cognitive science. See also [Bullock, 1994], [English, 1997], [Lakoff and Núñez, 1997], [Lakoff and Núñez, 1998], [Mac Lane, 1981], [Núñez, Edwards and Matos, 1999], [Núñez, 2000], [Núñez and Lakoff, 1998], [Pimm, 1988], [Sfard, 1994], [Sfard, 1997].

**minimize**    See maximize.

**minus**  The word **minus** can refer to both the binary operation on numbers, as in the expression $a - b$, and the unary operation of taking the **negative**: negating $b$ gives $-b$. In current usage in American high schools, $a - b$ would be pronounced "$a$ minus $b$", but $-b$ would be pronounced "negative $b$". The older usage for $-b$ was "minus $b$". College students are sometimes confused by this usage from older college teachers.

*Difficulties*  In ordinary English, if you **subtract** from a collection you make it smaller, and if you **add** to a collection you make it bigger. In mathematics, adding may also refer to applying the operation of addition; $a + b$ is smaller than $a$ if $b$ is negative. Similarly, subtracting $b$ from $a$ makes the result bigger if $b$ is negative. Both these usages occur in mathematical writing.

Students sometime assume that an expression of the form $-t$ must be negative. This may be because of the new trend of calling it "negative $t$", or because of the use of the phrase "opposite in **sign**". *Citations:* (305), (145), (290). *Reference:* [Hersh, 1997a].

**mnemonic**  A **mnemonic** identifier is one that suggests what it is naming.

*Example 1*  Mathematical mnemonic identifiers usually consist of the initial letter of the word the identifier suggests, as $f$ for a function, $G$ for a **group**, and so on.

*Example 2*  Category theorists use "**Ab**" for the category of Abelian groups.

See [Bagchi and Wells, 1998a]. See also **predicate symbol**. **suggestive name** and **tangent**.

*Acknowledgments:* Michael Barr.

## model

### 1. Model as mathematical object

In one of its uses in mathematical discourse, a **model**, or **mathematical model**, of a phenomenon is a **mathematical object** that represents the phenomenon. In fact, the mathematical object is often called a **representation** of the phenomenon. The phenomenon being modeled may be physical or another mathematical object.

**Example 1** A moving physical object has a location at each instant; this may be modeled by a **function**. One then observes that there is a relation between the derivative of the function and the average velocity of the physical object that allows one to define the instantaneous velocity of the object.

**Example 2** A "word problem" in algebra or calculus texts is an invitation to find a mathematical model of the problem (set it up as a mathematical **expression** using appropriate operations) and then solve for the appropriate variable. This example is of course closely related to Example 1.

**Example 3** Mathematical **logic** has a concept of model of a theory; a theory is a **mathematical object abstracting** the notion of structure subject to axioms. A model of the theory is then a set valued function that preserves both the mathematical and the inferential structure of the theory.

**Example 4** In computing science, a mathematical model of **algorithm** is defined. It may also be called an algorithm (but other names are used in some texts).

As the examples just discussed illustrate, a model and the thing it models are often called by the same name. Thus one refers to the velocity of an object (a physical property) and one also says that the derivative of the velocity is the acceleration. In fact a mathematical model is a special

*Models are to be used, not believed.*
—*Henri Theil*

*In theory, there is no difference between theory and practice. But in practice, there is.*
—*Jan van de Snepscheut*

kind of **metaphor** (see Example 3 under **metaphor**), and to refer to the mathematical model as if it were the thing modeled is a normal way of speaking about metaphors. It would be worth investigating whether and how this confuses students.

### 2. Physical model

A mathematical object may also have a **physical model**.

***Example 5*** A Möbius strip may be defined mathematically, and then modeled by taking a rectangular strip of paper, twisting it around halfway, and gluing the ends together.

*Citations:* (218), (275), (367), (411).

**modulo** The phrase "$x$ is the same as $y$ modulo $\sim$" means that $x$ and $y$ are elements of some set, $\sim$ is an equivalence relation on the set, and $x \sim y$. One would also say that $x$ is the same as $y$ **up to** $\sim$.

This is also used colloquially in phrases such as "The administration kept my salary the same modulo [or mod] inflation". Presumably the equivalence relation here is something like: "One dollar in 2002 is equivalent to \$1.02 in 2003."

In **number theory** one writes $a \equiv b \bmod n$, pronounced "$a$ is equivalent to $b \bmod n$", to mean that $a - b$ is divisible by $n$, for $a$, $b$, $n$ integers. Thus $20 \equiv 11 \bmod 3$. In this usage, the symbol "mod" occurs as part of a three-place assertion.

Many computer languages use an expression such as "$a \bmod n$" (the syntactic details may differ) to mean the remainder obtained when $a$ is divided by $n$; they would write $20 \bmod 3 = 2$. In this usage, mod is a **binary operation**.

Mathematics majors often enter a number theory class already familiar with the computer usage, to their resultant confusion. *Citations:* (75), (223), (222), (378)

**multiple meanings**  Some names and symbols in the mathematical register have more than one meaning.

*Example 1*  I recall as a graduate student being puzzled at the two meanings of **domain** that I then knew, with the result that I spent a (mercifully short) time trying to prove that the domain of a continuous function had to be a connected open set.

Following is a list of entries in this Handbook of words and symbols that have two or more distinct meanings. I have generally restricted this to cases where students are likely to meet both usages by the time they are first year graduate students in mathematics. See also **lazy evaluation**, **trigonometric functions** and **symbolic language**.

| | | |
|---|---|---|
| algebra 9 | field 97 | modulo 168 |
| algorithm 9 | formal 99 | or 184 |
| argument 19 | function 106 | order 185 |
| bracket 32 | graph 116 | parenthesis 192 |
| category 35 | I 120 | permutation 197 |
| composite 40 | identity 122 | power 202 |
| constructivism 52 | if 123 | prime 203 |
| contain 52 | image 125 | proposition 210 |
| continuous 54 | in general 126 | range 213 |
| continuum hypothesis 56 | injective 131 | result 219 |
| definition 66 | integral 133 | revise 219 |
| divide 76 | large 139 | sign 229 |
| domain 77 | logarithm 148 | subscript 235 |
| elementary 79 | map 150 | superscript 238 |
| equivalent 86 | minus 165 | tangent 247 |
| family 97 | model 167 | term 248 |

It is a general phenomenon that a particular phrase may mean different things in different branches of mathematics or science. Some of the words listed above fit this, for example **category**, **constructivism**, **continuum hypothesis**, **domain**, **field**, **graph**, **integral**. Of course, there are innumerable examples of others like this in mathematics or in any part of science. However, some of the words listed above can occur in both their meanings in the *same document*, for example contain, equivalent, identity, and the notations $(a, b)$ (see parenthesis) and $f^n(x)$ (see superscript).

**must**   One frequently finds "must be" used in the mathematical register when "is" would give the same meaning. It is used with verbs other than "be" in the same way. I presume this is to emphasize that the fact being asserted can be proved from facts known in the context.

Other uses of "must" in mathematical discourse are generally examples of the way the words is used in ordinary discourse.

**Example 1**   "If $m$ is a positive integer and $2^m - 1$ is prime, then $m$ must be prime."

**Example 2**   "Let $C = \{1, 2, 3\}$. If $C \subset A \cup B$, then one of $A$ and $B$ must contain two elements of $C$." *Citations:* (10), (199), (294), (295).

**myths**   Students in mathematics courses may have false beliefs about the subject which are perpetuated explicitly from class to class in their discussions with each other in attempting to explain a concept "in their own words". Some of the myths, sadly, are perpetuated by high school teachers. I list two here; it would be helpful to give them names as discussed under behaviors. Another example is given under element.

*(a) The empty set*   Many students in my discrete math classes frequently believe that the empty set is an element of every set. Readers of early versions of this book have told me that some high school teachers and even some college-level mathematicians believe this myth.

Other problems with the empty set are discussed in the entry about them.

*(b) Limits* The myth that a sequence with a limit "approaches the limit but never gets there" is discussed under limit.

See attitudes.

**N** The symbol ℕ usually denotes the set of natural numbers, in one or another of the meanings of that phrase. *Citations:* (124), (104).

**name** The **name** of a mathematical object is an English word or phrase used as an identifier of the object. It may be a determinate identifier or variate. It should be distinguished from a symbol used as an identifier. The distinction between name and symbol is discussed under identifier.

*Common words as names* A **suggestive name** is a a common English word or phrase, chosen to suggest its meaning. Thus it is a metaphor.

*Example 1* "Slope" (of a curve), or "connected subspace" (of a topological space). See the discussion of suggestive names in [Wells, 1995] and [Bagchi and Wells, 1998a]. See also tangent.

*Learned names* A name may be a new word coined from (usually) Greek or Latin roots. Such an identifier is a **learned name**. (Pronounce "learned" with two syllables.)

*Example 2* "Homomorphism".

*Personal names* A concept may be named after a person.

*Example 3* L'Hôpital's Rule, Hausdorff space.

*Typography* A mathematical object may be named by the typographical symbol(s) used to denote it. This is used both formally and in on-the-fly references. *Citations:* (44), (105), (414).

*Difficulties* The possible difficulties students may have with common words used as identifiers are discussed under formal analogy and semantic contamination. See also cognitive dissonance, names from other languages and multiple meanings.

*The author of the Iliad is either Homer or, if not Homer, somebody else of the same name.*
*–Aldous Huxley*

*References:* This discussion is drawn from [Bagchi and Wells, 1998b]. [Hersh, 1997a] gives many examples of dissonance between the mathematical meaning and the ordinary meaning of mathematical words.

**namely**   Used to indicate that what follows is an explication (often a repetition of the definition) of what precedes.

**Example 1**   "Let $G$ be an Abelian group, namely a group whose multiplication is commutative."

**Example 2**   "We now consider a specific group, namely $S_3$".

The word is also used after an existence claim to list those things that are claimed to exist. (Of course, this is a special case of explication.)

**Example 3**   "12 has two prime factors, namely 2 and 3."
    *Citations:* (271), (272), (297), (309). (383).

**names from other languages**   Mathematicians from many countries are mentioned in **mathematical discourse**, commonly to give them credit for theorems or to use their **names** for a type of **mathematical object**. Two problems for the student arise: Pronunciation and variant spellings.

**(a) Pronunciation**   During the twentieth century, it gradually became an almost universal attitude among educated people in the USA to stigmatize pronunciations of words from common European languages that are not approximately like the pronunciation in the language they came from, **modulo** the phonologies of the other language and English. This did not affect the most commonly-used words. The older practice was to pronounce a name as if it were English, following the rules of English pronunciation.

For example, today many mathematicians pronounce "Lagrange" the French way, and others, including (in my limited observation) most engineers, pronounce it as if it were an English word, so that the second syllable rhymes with "range". I have heard people who used the second pronunciation corrected by people who used the first (this happened to me when I was a graduate student), but never the reverse when Americans are involved.

Forty years ago nearly all Ph.D. students had to show mastery of two foreign languages; this included pronunciation, although that was not emphasized. Today the language requirements in the USA are much weaker, and educated Americans are generally weak in foreign languages. As a result, graduate students pronounce foreign names in a variety of ways, some of which attract ridicule from older mathematicians. (Example: the possibly apocryphal graduate student at a blackboard who came to the last step of a long proof and announced, "Viola!", much to the hilarity of his listeners.) There are resources on the internet that allow one to look up the pronunciation of common foreign names; these may be found on the website of this handbook.

> The older practice of pronunciation evolved with the English language: In 1100 AD, the rules of pronunciation of English, German and French, in particular, were remarkably similar. Over the centuries, the sound systems changed, and Englishmen, for example, changed their pronunciation of "Lagrange" so that the second syllable rhymes with "range", whereas the French changed it so that the second vowel is nasalized (and the "n" is not otherwise pronounced) and rhymes with the "a" in "father".

**(b) Transliterations from Cyrillic** Another problem faced by the mathematics graduate student is the many ways of transliterating foreign names. For example, name of the Russian mathematician most commonly spelled "Chebyshev" in English is also spelled Chebyshov, Chebishev, Chebysheff, Tschebischeff, Tschebyshev, Tschebyscheff and Tschebyschef. (Also Tschebyschew in papers written in German.) The *correct* spelling of his name is Чебышев, since he was Russian and Russian used the Cyrillic alphabet.

> In spite of the fact that most of the transliterations of "Чебышев " show the last vowel to be an "e", the name in Russian is pronounced approximately "chebby-SHOFF", accent on the last syllable.

The only spelling in the list above that could be said to have some official sanction is Chebyshev, which is used by the Library of Congress. This is discussed by Philip J. Davis in [1983]. Other citations: (127), (136), (151), (158), (197), (366).

*I would rather decline
two German beers than
one German adjective.*
  *−Mark Twain*

*(c) German spelling and pronunciation* The German letters "ä", "ö" and "ü" may also be spelled "ae", "oe" and "ue" respectively. The letters "ä", "ö" and "ü" are alphabetized in German documents as if they were spelled "ae", "oe" and "ue". It is far better to spell "Möbius" as "Moebius" than to spell it "Mobius".

The letter "ö" represents a vowel that does not exist in English; it is roughly the vowel sound in "fed" spoken with pursed lips. It is sometimes incorrectly pronounced like the vowel in "code" or in "herd". Similar remarks apply to "ü", which is "ee" with pursed lips. The letter "ä" may be pronounced like the vowel in "fed".

The German letter "ß" may be spelled "ss" and often is by Swiss Germans. Karl Weierstrass spelled his last name "Weierstraß". Students sometimes confuse the letter "ß" with "f" or "r". In English language documents it is probably better to use "ss" than "ß".

Another pronunciation problem that students run into are the combinations "ie" and "ei". The first is pronounced like the vowel in "reed" and the second like the vowel in "ride". Thus "Riemann" is pronounced REE-mon.

**narrative style**   The **narrative style** of writing mathematics is a style involving infrequent labeling; most commonly, the only things labeled are **definitions**, **theorems**, **proofs**, and major subsections a few paragraphs to a few pages in length. The reader must deduce the logical status of each sentence from connecting phrases and bridge sentences. This is the way most formal mathematical prose is written.

*Difficulties* Students have difficulties of several types with narrative proofs.
- The proof may leave out steps.
- The proof may leave out reasons for steps.

- The proof may instruct the reader to perform a calculation which may not be particularly easy. See **proof by instruction**.
- The proof may not describe its own structure, which must be determined by **pattern recognition**. See **proof by contradiction**.
- The proof may end without stating the conclusion; the reader is expected to understand that the last sentence of the proof implies the conclusion of the theorem via known facts. Example 5 under **pattern recognition** gives a proof that two sides of a triangle are equal that ends with "Then triangle $ABC$ is congruent to triangle $ACB$ ... "; the reader must then see that the congruence of these two triangles implies that the required sides are the same.

Contrast **labeled style**. *References:* This style is named and discussed in [Bagchi and Wells, 1998a]. See [Selden and Selden, 1999].

**natural number**   For some authors, a **natural number** is a **positive integer**. For others it is a nonnegative **integer**, and for others it is any integer. It appears to me that the most common meaning these days is that a natural number is a nonnegative integer. *Citations:* (124), (104), (191), (311).

***Remark 1*** As the **citations** show, the disagreement concerning the meaning of this phrase dates back to the nineteenth century.

**necessary**   $Q$ is **necessary** for $P$ if $P$ **implies** $Q$. Examples are given under **conditional assertion**.

The motivation for the word "necessary" is that the **assertion** "$P$ implies $Q$" is logically **equivalent** to "not $Q$ implies not $P$" (see **contrapositive**), so that for $P$ to be true it is necessary in the usual sense of the word for $Q$ to be true.

**negation**   The **negation** of an **assertion** $P$ is an assertion that denies $P$. In some circumstances that is the effect of the English word **not**. In symbols, "not $P$" may be written $\neg P$, $-P$, or $\bar{P}$. *Citation:* (383).

**(a) *Negation of quantified statements*** If $P(x)$ is a **predicate** possibly containing the **variable** $x$, then the negation of the **assertion** $\forall x\, P(x)$ is $\exists x\, \neg P(x)$. Similarly, the negation of the assertion $\exists x\, P(x)$ is $(\forall x)\neg P(x)$.

Both of these rules cause difficulty in translating to and from English. It is my experience that many students need to be explicitly taught these rules and how to express them in English.

**Example 1** The negation of the **assertion**
"All multiples of 4 are even."
is not
"All multiples of 4 are not even."
but rather
"Multiples of 4 are not all even."
or, equivalently,
"Not all multiples of 4 are even."

This illustrates the fact that simply putting a "not" into a sentence may very well give the wrong results.

**Example 2** In colloquial English as spoken by many people (including students!), the sentence
"All multiples of 3 are not odd."
means that *some* multiples of 3 are not odd (a true statement). A similar remark holds for "Every multiple of 3 is not odd." I believe that most mathematicians would interpret it as meaning that no multiple of 3 is odd (a false statement). See **esilism**.

This phenomenon quite possibly interferes with students' understanding of negating quantifiers, but I have no evidence of this.

"Negation" is also used sometimes to denote the operation of taking the **negative** of a number. *Citations:* (297).

KNOT.

NOT.

A KNOT IS NOT A NOT.

**negative**   See minus.

**never**   An assertion about a variable mathematical object of the form "$A$ is never $B$" means that for all $A$, $A$ is not $B$. An assertion of that form when $A$ is a function means that no value of $A$ is $B$.

*Example 1*   "A real number never has a negative square."

*Example 2*   "The sine function is never greater than 1."

> See also always, time and universal quantifier
> *Citations:* (114), (199), (331).

**not**   See negation.

**nonnegative**   A real number $r$ is **nonnegative** if it is not negative, in other words if $r \geq 0$. *Citations:* (124), (282).

*Remark 1*   In an ordered set containing an element denoted 0, the statement $r \geq 0$ and the statement "not $r < 0$" are not equivalent if the set is not totally ordered.

**notation**   Notation is a system of signs and symbols used as a representation of something not belonging to a natural language. The symbolic language of mathematics is a system of notation. See establish notation.

**noun phrase**   A **noun phrase** in English consists of these constituents in order: a **determiner**, some **modifiers**, a noun called the **head** of the noun phrase, and some more modifiers. The only thing in this list that *must* be there is the noun. A noun phrase typically describes or names something; thus it is a phrase that acts as a noun.

The determiner may be an **article** or one of a small number of other words such as "this", "that", "some", and so on. The modifiers may be adjectives or certain types of phrases or clauses.

A noun phrase may occupy one of a number of different **syntactic** positions in a **sentence**, such as the subject, direct or indirect object, predicate nominative, object of a preposition, and so on.

**Example 1**   This is a noun phrase:

$$\underbrace{\text{The}}_{Det} \underbrace{\text{little brown}}_{Modifiers} \underbrace{\text{fox}}_{Noun} \underbrace{\text{in the bushes}}_{Modifier}$$

In this noun phrase, the head is "fox". The word "bushes" is also a noun but it is not the head; it is a constituent of a modifying phrase.

This description omits many subtleties. See **definite description** and **indefinite description**.

*Reference:* [Greenbaum, 1996].

## now

**(a) *Introduce new notation***   "Now" may indicate that new notation or assumptions are about to be introduced. This is often used to begin a new **argument**. This use may have the effect of canceling assumptions made in the preceding text. *Citations:* (4), (94), (109), (200), (285).

**Example 1**   "We have shown that if $x \in A$, then $x \in B$. Now suppose $x \in B$. ... "

**(b) *Bring up a fact that is needed***   "Now" may be used to point out a fact that is already known or easily deduced and that will be used in the next step of the **proof**.

**Example 2**   In a situation where we already know that $x = 7$, one could say:

    " ... We get that $x^2 + y^2 = 100$. Now, $x$ is 7, so $y = \sqrt{51}$."

This is similar to the second meaning of **but**.

*Citation:* (16).

*(c) "Here"* "Now" may simply refer to the point in the text at which it occurs. The **metaphor** here is that the reader has been reading straight through the text (unlike a **grasshopper**) and at the current moment she sees this word "now". As such it does not really add anything to the meaning. *Citations:* (328), (360).

**Remark 1** The three usages described here are not always easy to distinguish.

*Acknowledgments:* Atish Bagchi

**number** Numbers in mathematics are usually written in base-10 notation, although most students these days are familiar with other bases, particularly 2, 8 and 16.

*(a) Type of number* The word **number** in most mathematical writing is used for one of the **types natural number** (whatever it means), **integer**, rational number, **real number** or complex number. A piece of discourse will commonly establish right away how the word is being used. If it does not, it is commonly because the type is clear from context, for example because it is being used to refer to the size of a collection. Occasionally, the context does not immediately make the usage clear, though further reading usually determines it. In the **citations** I have found where the context does not immediately make the usage clear, it always turns out to be real. *Citations:* (114), (211), (239).

*(b) Variables and base-10 notation* The syntax for the usual base-10 notation and for variables of type **integer** (or other type of number) are different in ways that sometimes confuse students.

**Example 1** Any nonzero number in base-10 notation with a **minus sign** in front of it is negative. This may cause the student to assume that the number represented by $-n$ (for example) is negative. Of course, it need not be, for example if $n = -3$.

***Example 2*** If $x$ and $y$ are numerical **variables**, $xy$ is their product. However, the **juxtaposition** of numbers in base-10 notation does *not represent* their product: 32 is not the product of 3 times 2. This causes some difficulty, perhaps mostly in high school. It has been said that college students sometimes cancel the $x$ in expressions such as

$$\frac{\sin x}{\cos x}$$

but I have not met up with this phenomenon myself.

See also item (i) under **behaviors**.

**number theory** The phrase **number theory** refers to the study of the **integers**, particularly with respect to properties of prime numbers. *Citations:* (87), (191), (382).

**object** See APOS and mathematical object.

**object-process duality** Mathematicians thinking about a mathematical concept will typically hold it in mind both as a **process** and an **object**. As a process, it is a way of performing mathematical actions in stages. But this process can then be conceived as a **mathematical object**, capable for example of being an element of a set or the input to another process. Thus the sine function, like any function, is a process that associates to each number another number, but it is also an object which you may be able to differentiate and integrate.

The mental operation that consists of conceiving of a process as an object is called **encapsulation**, or sometimes **reification** or **entification**. Encapsulation is not a one-way process: while solving a problem you may think of for example finding the antiderivative of the sine function, but you are always free to then consider both the sine function and its antiderivative as processes which can give values –

This discussion makes it sound as if the mathematician switches back and forth between process and object. In my own introspected experience, it is more like holding both conceptions in my mind at the same time. See [Piere and Kieren, 1989].

180

and then you can conceive of them encapsulated in another way as a graph in the $xy$ plane.

The word **procept** was introduced in [Gray and Tall, 1994] to denote a **mathematical object** together with one or more processes, each with an **expression** that encapsulates the process and simultaneously denotes the object. Thus a mathematician may have a procept including the number 6, expressions such as $2+3$ and $2 \cdot 3$ that denote calculations that result in 6, and perhaps alternative representations such as 110 (binary). This is similar to the idea of **schema**. See also **APOS** and **semantics**.

*References:* [Gray and Tall, 1994], [Sfard, 1991], [Sfard, 1992] (who gives a basic discussion of mathematical objects in the context of functions), [Carlson, 1998], [Dubinsky and Harel, 1992], [Hersh, 1997b], pages 77ff, [Thompson and Sfard, 1998].

**obtain**   Most commonly, "obtain" means "get", as in ordinary English.
***Example 1***   "Set $x = 7$ in $x^2 + y^2 = 100$ and we obtain $y = \sqrt{51}$."

*Citations:* (33), (147).

In the **mathematical register**, "obtain" may also be used in much the same way as **hold**. This usage appears uncommon.
***Example 2***   "Let $G$ be a group in which $g^2 = e$ obtains for every element $g$."

*Citations:* (153), (428).

*Acknowledgments:* Atish Bagchi.

## on

### 1. Function on domain
A function $F$ is **on** a set $A$, or **defined in**, **defined on** or defined over $A$, if its **domain** is $A$. *Citations:* (54), (84), (403).

### 2. Structure on underlying set
A structure is **on** $A$ if its **underlying set** is $A$. *Citation:* (128).

### 3. Set where condition is satisfied

A mathematical structure is "$P$ on $A$", where $P$ is some **condition**, if the structure has a **parameter** that varies over some set **containing** $A$, and $P$ is true of the structure if the parameter is in $A$.

**Example 1**   "$x^2 - 1$ is positive on the interval $[2, 3]$." *Citations:* (143), (275).

### 4. On a figure

A point $p$ is **on** a set if it is in the set. This usage seems to be restricted to geometric figures.

**Example 2**   "The point $(\frac{1}{\sqrt{2}}, \frac{1}{\sqrt{2}})$ is on the unit circle." *Citations:* (155), (172).

**one to one**   Injective.

**only if**   In the mathematical register, if $P$ and $Q$ are **assertions**, "$P$ only if $Q$" means $P$ **implies** $Q$. The phrase "only if" is rarely used this way in ordinary English discourse.

**Example 1**   The sentence
    "4 divides $n$ only if 2 divides $n$"
means the same thing as the sentence
    "If 4 divides $n$, then 2 divides $n$."

**Example 2**   The sentence
    "I will carry my umbrella only if it rains."
does not mean the same thing as
    "If I carry my umbrella, it will rain."

***Difficulties*** Students often get the sentence in Example (1) backward, taking it to mean

$$(2 \text{ divides } n) \Rightarrow (4 \text{ divides } n)$$

Some of them flatly refuse to believe me when I tell them the correct interpretation. This is a classic example of **semantic contamination**, a form of **cognitive dissonance** – two sources of information appear to contradict each other, in this case the professor and a lifetime of intimate experience with the English language, with the consequence that one of them is rejected or suppressed. It is hardly suprising that some students prefer to suppress the professor's apparently unnatural and usually unmotivated claims.

McCawley [1993] also rejects the equivalence of "*A* only if *B*" with "If *A*, then *B*", for ordinary **discourse**, but in the **mathematical register** the sentence must be taken to be equivalent to the others. This difference may have come about because **conditional assertions** in ordinary English carry connotations of *causality* and *time dependence*. Because **mathematical objects** are thought of as **inert** and **eternal**, the considerations that distinguish the two sentences in the example do not apply to statements such as the sentence in Example (1); the truth of the statement is determined entirely by the truth table for **implication**.

The remarks in the preceding paragraph may explain some of the difficulties students have with the **contrapositive**, as well.

**onto**   Surjective.

**operation**   Used to refer to a **function of two variables** that is written in **infix notation**. May be called a **binary operation**

***Example 1***   The operation of addition on the set of real numbers is a binary operation.

*Citation:* (273). Some authors use "operation" in certain contexts to refer to any function. *Citation:* (385).

**operator**   Operator means function. Most authors seem to use "operator" only in certain restricted situations. It is often used when the domain is a set of functions or when the operator is a function from a **space** to itself. *Citation:* (72). (But a "linear operator" can be between different spaces.)

The text [Grassman and Tremblay, 1996] uses "operator" to refer to a **binary operation** used in **infix notation** (see the discussion on pages 104 through 108). The text [Gries and Schneider, 1993] takes a similar approach (page 7 and page 387). The word is used to refer both to the **symbol** and to the **function**. This usage may be associated with authors having a background in computing science or logic.

*Acknowledgments:* Atish Bagchi and Michael Barr.

**or**   Or placed between two **assertions** produces the **disjunction** of the assertions.

**Example 1**   "$x$ is nonnegative or $x < 0$".

**Terminology**   In **mathematical logic**, "or" may be denoted by "∨" or "+".

**Difficulties**   As the truth table for **disjunction** indicates, "$P$ or $Q$" allows both $P$ and $Q$ to be **true**, although they cannot both be true in the example just given. The assertion

" $x > 0$ or $x < 2$ "

is *true* for any real number $x$. A student may feel discomfort at this assertion, perhaps because in many assertions in conversational English involving "or" both cases cannot happen. Authors often emphasize the inclusiveness by saying something such as "or both".

See [Hersh, 1997a] for more examples. *Citations:* (75), (91), (302).

Students also have trouble negating **conjunctions** and disjunctions. A statement such as

"$x$ is not $(P$ or $Q)$"

means

"$x$ is not $P$ *and* $x$ is not $Q$."

So does

"$x$ is neither $P$ nor $Q$."

See also **both** and Example 3 under **yes it's weird**.

**or equivalently**    This phrase means that what follows is **equivalent** to what precedes. It is usually used when the equivalence is easy to see. This usage has no relation to the **connective** "or". *Citation:* (47).

## order

### 1. Ordering
"Order" may be a variant of "ordering".

**Example 1**    "Let $\leq$ be the usual order on the real numbers." *Citations:* (104), (242).

### 2. Cardinality
The **order** of a structure such as a **group** is the **cardinality** of (the un-derlying set of) the structure. *Citations:* (369). But the meaning can be more devious than that: See (250).

### 3. Parameter
The word "order" may refer to a nonnegative integer **parameter** or func-tion of the structure. Of course, the cardinality meaning just mentioned is a special case of this.

**Example 2**    The **order** of a differential equation is the highest deriva-tive occurring in the equation.

**Remark 1**    The word **degree** is also used in this way, but the uses are not interchangeable. Indeed, a structure may have both an order and a degree, for example a permutation **group**. *Citations:* (192), (356).

*In New York City a man is mugged every eleven seconds. And this is that man!*
  *–Robin Williams with John Belushi on Saturday Night Live.*

No dependence is suggested if the two quantifiers that occur in order are the same. Thus $\forall x \forall y P(x, y)$ means the same as $\forall y \forall x P(x, y)$ and $\exists x \exists y P(x, y)$ means the same as $\exists x \exists y P(x, y)$.

**order of quantifiers**   When two quantifiers occur one after the other in a mathematical statement, a dependence between the **variables** they **bind** may be suggested, depending on the order of occurrence of the quantifiers.

**Example 1**   The statement

$$\forall x \exists y (xy = e)$$

about elements $x$ and $y$ of a **group** with identity $e$, says that every element has a right inverse; that is satisfied by all groups. In contrast, the **statement**

$$\exists x \forall y (xy = e)$$

is satisfied only by the trivial group. The idea is that the element $y$ in the first sentence depends on the element $x$, and that according to the customary interpretation of sentences in **mathematical logic**, this is signaled by the fact that the $x$ comes first. (See **esilize** for more about this example.)

**Example 2**   The **definition** of **continuity** commonly begins this way:

  "For every $\epsilon > 0$, there is a $\delta > 0$ for which..."

Here $\delta$ depends on $\epsilon$, but in contrast to the preceding example, the dependence is not functional.

  See [Bagchi and Wells, 1998b].

**Difficulties**   Reversing the definition of continuous to write "There is a $\delta > 0$ such that for every $\epsilon > 0$" gives the definition of uniform continuity. Mathematicians in the nineteenth century had a great deal of difficulty separating these two ideas, so it is hardly surprising that our students do, too.

  In ordinary English the way quantifiers are ordered does not always obey these rules. A student might say, "there is an inverse for every element" and be understood in much the same way as one would understand a statement such as "there is an ice cream cone for every child". The latter statement, translated mindlessly into first order logic, brings up the picture of $n$ children licking one cone. But no one in everyday discourse would understand it that way, and only a few **esilists** would think it bad English. Nevertheless, in writing mathematical **arguments** in English, such constructions are avoided by many authors (see **esilize**).

It appears to me that the meaning of sentences such as "There is an ice cream cone for every child" is extracted using a mechanism similar to that for a **distributive plural**, but I have not found anything in the linguistics literature about this.

See also **all**, **and**, **mathematical logic** and **writing dilemma**. *Reference:* [Dubinsky, 1997].

**orthogonal**   A system of **notation** is **orthogonal** if any construction possible in the notation can be used anywhere it is appropriate.

***Example 1***   The notation for derivatives is not orthogonal. The prime notation can be used for **functions** of one **variable** but not for functions of more than one variable. Thus the failure of orthogonality occurs because the prime notation **suppresses a parameter** (the variable with respect to which the derivative is taken).

***Example 2***   The notation involving $d$ is used for functions of one variable; for more than one variable one must change it to $\partial$. (There is a sense in which this is not a failure of orthogonality, although students generally are not aware of this. See Example 3 under **irregular syntax** for a discussion.)

***Example 3***   Early forms of Fortran were not orthogonal; one could use an arithmetic expression (for example, $i+2j^2$) that evaluated to an integer in most places where one could use an integer — but not in the **subscript** of an array. In other words, one could write the equivalent of $A_i$ but not of $A_{i+2j^2}$. This context is where I first met the word "orthogonal" used. *Citations:* (18), (324).

Of course, "orthogonal" has a meaning as a technical word in mathematics, as well.

**osmosis theory**   The **osmosis theory** of teaching is this **attitude**: We should not have to teach students to understand the way mathematics is written, or the finer points of logic (for example how **quantifiers** are **negated**. They should be able to figure these things on their own — "learn it by osmosis". If they cannot do that they are not qualified to major in mathematics. (See **mathematical mind**).

We learned our native language(s) as children by osmosis. That does not imply that **college** students can or should learn mathematical reasoning that way. It does not even mean that **college** students should learn a foreign language that way.

**outfix notation**   A function is displayed in **outfix notation** if its **symbol** consists of **characters** or **expressions** put on both sides of the **argument**.

*Example 1*   The absolute value of a number $r$ is denoted $|r|$.

*Example 2*   The greatest integer in $x$ is sometimes denoted by $\lfloor x \rfloor$.

Other examples are described under **brace**, **angle bracket** and **bracket**. See also **integral**. Also called **matchfix notation** *Citations:* (12).

**output**   See **function**.

**over**

*(a) Specifying domain*   To say that a function $f$ is **defined over** $S$ means that the **domain** of $f$ is $S$. *Citation:* (349).

*(b) Specifying a property with respect to a substructure* Some objects come attached with parameters, and one can say that the object does or does not have some property **over** that parameter.

*Example 1* The polynomial $x^2+1$ does not split over the real field, but it does split over the complex number field. Here the field is a parameter for the polynomial.

*(c) Defining an associated structure with respect to a substructure* One also defines associated structures in terms of the object and the parameter.

*Example 2* $\int_a^b x^3\,dx$ is positive over any interval $[a,b]$ in the positive half interval. *Citations:* (229), (314).

*(d) Specifying a structure with projection to a given structure* Many **mathematical structures** have as part of their definition a projection onto some other structure $S$, not necessarily of the same kind. Thus one talks about a sheaf **over** a space, or a multisheeted surface over a region in the complex plane. Such a projection will also define an inverse image structure **over** a point in $S$. *Citations:* (405) (257).

Here the word **over** carries a metaphor of *literally* "over", since such structures are typically drawn with the projection going down. See **law of gravity for functions**. *Citations:* (50), (177).

**overloaded notation** This phrase usually applies to a **symbol** or a name for a **function** that takes on different meanings depending on which **type** of element it is evaluated at. Such a function is also called **polymorphic**.

*Example 1* The **identity function** is a polymorphic name; in the usual formalism there is a different identity function on each set.

***Example 2***   A familiar example is the **symbol** $\times$, which is overloaded in **college mathematics** courses. When $a$ and $b$ are numbers, $a \times b$ is their product. When $A$ and $B$ are matrices, $A \times B$ is the matrix product. When $v$ and $w$ are 3-**vectors**, $v \times w$ is their vector product.

***Example 3***   Another example is the common treatment of the **image** for arbitrary functions: Let $F : S \to T$ be a **function**.

a) If $x \in S$, $F(x)$ is the value of $F$ applied to $x$. It is called the **image of $x$ under** $F$.

b) If $A$ is a subset of $S$, then $F(A) = \{F(x) \mid x \in A\}$ (see **setbuilder notation**). It is called the **image of $A$ under** $F$.

c) The image of $F$ is the set of all $t$ in $T$ for which there is an $x \in S$ such that $F(x) = t$, which is the image in the sense of (b) of the **domain** of $F$. The word "**range**" is also used for this meaning.

***Remark 1***   The preceding example is in a way fake. One could simply stipulate that *every* function $F : S \to T$ has values defined for every element of $S$ and (in the way illustrated above) for every subset of $S$. However, the phrase "the image of $F$" would still overload the word "image".

When students begin **college mathematics**, the frequent occurrence of ad-hoc polymorphism means that they have to *read the surrounding text* to understand what a **symbolic expression** means: the expression is no longer self-sufficient. When I first came across this aspect of mathematics in a matrix theory course at Texas Southmost College, I felt that I had been ejected from paradise.

***Example 4***   A functor $F$ from a category $\mathcal{C}$ to a category $\mathcal{D}$ is defined on both objects and arrows of $\mathcal{C}$. This, too is a fake example, since the value of the functor at identity arrows determines its value on objects.

***Example 5***   A text on vector spaces will very likely use $+$ for addition of **vectors** in every vector space. Similarly, some texts on **group** theory will use $e$ or $1$ for the identity element for any group and juxtaposition for the **binary operation** of any group.

***Remark 2***   Example 5 illustrates the common case of using the same **symbol** in every model for a particular operation in an axiomatically defined **mathematical structure**.

***Remark 3*** The operation $\times$ does not require the same algorithm on matrices as it does on 3-vectors. This is the sort of phenomenon computer scientists call **ad-hoc polymorphism**. It is contrasted with **parametric polymorphism**, an example of which is the algorithm "swap the two entries in an ordered pair", which applies to ordered pairs of any type of element. (The parameter that gives rise to the name "parametric" is the type of element.) See algorithm. The identity function provides a trivial example of parametric polymorphism.

Many mathematicians think and speak informally of a parametrically polymorphic function as one single function. (" ... the identity function is injective").

***Remark 4*** The concept "overloaded" is natural in computing science because operations on different data types are typically implemented differently. For example, addition of integers is implemented differently from addition of floating point numbers in most computer languages. The concept is less natural in mathematics, where you could define the operation on the disjoint union of all the sets under consideration (for $\times$, the set might be $\mathbb{R}$ plus the set of all 3-dimensional real vectors plus the set of all $n \times n$ real matrices for each $n$). Even there, however, the implementation algorithm differs for some of the subsets. (See cases.)

See also superscript.

*Citation:* (248).

**parameter**  A **parameter** is a variate identifier used in the definition of a mathematical object. When the parameters are all instantiated, the object becomes specific. The parameters may or may not be shown explicitly in the identifier for the object; see synecdoche and suppression of parameters. See also Example 2 under definite description.

***Example 1***  Let $[a, b]$ be a closed interval. Here the parameters are $a$ and $b$. A particular instantiation gives the specific closed interval $[\pi, 2\pi]$.

***Example 2***   Consider the polynomial $x^2 + ax + b$. The parameters are again $a$ and $b$. (See Remark 1 below.)

***Example 3***   Consider the function $f(x) = x^2 + ax + b$. Again $a$ and $b$ are parameters and $x$ is not. This might typically be referred to as a "two-parameter family $f_{a,b}(x) = x^2 + ax + b$ of functions".

***Remark 1***   A parameter in a **symbolic expression** is necessarily a **free variable**, but the **converse** may not be true.

In Example 3, the variable $x$ is not free, because the definition of $f$ uses a **defining expression** based on the named input variable $x$ (see **bound variable**). In consequence, the only parameters are $a$ and $b$.

On the other hand, in the expression $x^2 + ax + b$ mentioned in Example 2, for example, all the **variables** $a$, $b$ and $x$ might be considered free, but one might refer only to $a$ and $b$ as parameters. In fact, whether $x$ is free or not depends on your point of view. If you think of $x^2 + ax + b$ as an expression to be **evaluated**, you must substitute for all three variables to get a number. But if you refer to $x^2 + ax + b$ as a polynomial as I did in the example, then **convention** decrees that it is a quadratic polyonial in $x$ with parameters $a$ and $b$ (because $x$ is near the end of the alphabet). In that case, substituting a number for $x$ destroys the fact that it is a polynomial, so there is an **argument** that the $x$ is at least psychologically *not* free.

Observe that in this remark I am saying that which variables are regarded as parameters is determined by *linguistic usage* and point of view, not by *mathematical definitions*.

*Citations:* (49), (252), (308), (321).

People outside science, particularly businessmen, often use "parameter" to mean "boundary", presumably because they confused the word with "perimeter".

**parenthesis**   Parentheses are the **symbols** "(" and ")". Parentheses are used in various ways in **expressions**.

*(a) Grouping*  Parentheses are very commonly used as **bare delimiters** to group **subexpressions**.

*Example 1*  Parentheses are used for grouping in the **expression** such as $(x^2 + 1)^2$ and $x(y + z)$. *Citations:* (103), (36), (427).

*(b) Tuples and matrices*  Parentheses may be used to denote an ordered *n*-tuple, as in $(3, 1, 2)$, and are used in the standard notation for matrices. *Citations:* (33), (36), (240)

*(c) Open interval*  The symbol $(a, b)$ may denote the real **interval**
$$\{x \mid a < x < b\}$$
*Citation:* (140).

*(d) Greatest common divisor*  The symbol $(m, n)$ may denote the greatest common divisor of the integers $m$ and $n$. *Citation:* (140).

The **citations** for this and the last usage come from the *same sentence*, which I quote here for convenience:

> Richard Darst and Gerald Taylor investigated the differentiability of functions $f^p$ (which for our purposes we will restrict to $(0, 1)$) defined for each $p \geq 1$ by
> $$f(x) = \begin{cases} 0 & \text{if } x \text{ is irrational} \\ 1/n^p & \text{if } x = m/n \text{ with } (m, n) = 1. \end{cases}$$

It appears to me quite unlikely that any experienced mathematician would be confused by that sentence. Students are another matter.

*(e) Function values*  It is not clear whether the use of **parentheses** to delimit the **arguments** in denoting the **value** of a function, in for example $f(x + 1)$, is a simple matter of grouping, or whether it is part of a special syntax for function application. See **irregular syntax**.

(HELP)

***Terminology*** Parentheses are also called brackets, but "bracket" may also refer to other delimiters. Sometimes parentheses are called **round parentheses** for emphasis.

**parenthetic assertion**    A symbolic assertion is **parenthetic** if it is embedded in a sentence in a natural language in such a way that its pronunciation becomes a phrase (not a clause) embedded in the sentence. In contrast, when a symbolic assertion is a clause it is pronounced in such a way as to be a complete sentence.

***Example 1***    "For any $x > 0$ there is a $y > 0$ such that $y < x$." The assertion "$x > 0$" in isolation is a complete sentence, typically pronounced "$x$ is greater than 0". In the sentence quoted above, however, it is pronounced "$x$ greater than 0" or "$x$ that is greater than 0", becoming a noun phrase embedded in the main sentence. Note that in the quoted sentence, "$x > 0$" and "$y > 0$" are parenthetic but "$y < x$" is a full clause. *Citations:* (28); (56); (112); (161); (335); (432).

***Remark 1***    In seeking citations I was struck by the fact that some authors use parenthetic assertions in almost every paragraph and others essentially never do this: the latter typically use symbolic assertions only as complete clauses. Compare the articles [Bartle, 1996] and [Neidinger and Annen III, 1996], in the same issue of The American Mathematical Monthly.

***Example 2***    "...we define a *null set* in $I := [a, b]$ to be a set that can be covered by a countable union of intervals with arbitrarily small total length." This is from [Bartle, 1996], page 631. It could be read in this way: "...we define a null set in $I$, which is defined to be of the form $[a, b]$, to be a set...". In other words, the phrase "$I := [a, b]$" is a definition *occurring as a parenthetic assertion.*

***Example 3***    "Consider the circle $S^1 \subseteq \mathbf{C} = \mathbf{R}^2$" This example is adapted from [Zulli, 1996]. Notice that the parenthetic remark contains another

194

parenthetic remark inside it.

> See also **context-sensitive**.

> *References:* [Gillman, 1987], pages 12–13; [Krantz, 1997], page 25; [Lamport, 1994], page 18.

> See **writing dilemma**.

**partition**   See **equivalence relation**.

**pathological**   See **example**.

**pattern recognition**   Mathematicians must recognize abstract patterns that occur in symbolic expressions, geometric figures, and in their own mental representations of **mathematical objects**. This is one small aspect of human pattern recognition; for a general view, see [Guenther, 1998], Chapter 3.

*(a) An expression as an instance of substitution*   One particular type of pattern recognition that students find immensely difficult is *recognizing that a given* **expression** *is an instance of a* **substitution** *into a known expression.*

**Example 1**   This Handbook's **definition** of "**at most**" says that "$x$ is at most $y$" means $x \leq y$. To understand this definition requires recognizing the **pattern** "$x$ is at most $y$" no matter what occurs in place of $x$ and $y$. For example,

> "$\sin x$ is at most 1"

means that $\sin x \leq 1$.

**Example 2**   Students may be baffled when a **proof** uses the fact that $2^n + 2^n = 2^{n+1}$ for positive integers $n$. This requires the recognition of the pattern $x + x = 2x$. Similarly $3^n + 3^n + 3^n = 3^{n+1}$.

*Art is the imposing of a pattern on experience, and our aesthetic enjoyment is recognition of the pattern.*
*Alfred North Whitehead*

***Example 3***  The assertion
$$"x^2 + y^2 > 0"$$
has as a special case
$$"(-x^2 - y^2)^2 + (y^2 - x^2)^2 > 0."$$
where you must insert appropriate parentheses. Students have trouble with expressions such as this one not only in recognizing it as an instance of substitution but in performing the substitution in the first place (see **substitution**).

*(b) Recognizing patterns of proof*  Students in postcalculus courses must recognize patterns of proof without being told. Examples are given under **contrapositive** and **proof by contradiction**.

> One of my favorite things is to recognize hidden patterns, in mathematics (for example how the properties of models reflect the structure of their theories) or in everyday experience (driving around northern Ohio and recognizing the former beaches left by the glaciers).

Some **proofs** involve recognizing that a **symbolic expression** or figure fits a pattern *in two different ways*. This is illustrated by the next two examples. I have seen students flummoxed by Example 4, and Example 5 may for all I know be the proof that flummoxed medieval geometry students (see **pons asinorum**).

***Example 4***  In set with an associative **binary operation** and an identity element $e$, suppose $x$ is an element with an inverse $x^{-1}$. (In this situation, it is easy to see that $x$ has only one inverse.)

Theorem: $(x^{-1})^{-1} = x$.

Proof: By definition of inverse, $y$ is the inverse of $x$ if and only if
$$xy = yx = e \tag{2}$$
It follows by putting $x^{-1}$ for $x$ and $x$ for $y$ in Equation (2) that we must show that
$$x^{-1}x = xx^{-1} = e \tag{3}$$
But this is true by that same Equation (2), putting $x$ for $x$ and $x^{-1}$ for $y$.

196

***Example 5*** Theorem: If a triangle has two equal angles, then it has two equal sides.

Proof: In the figure below, assume $\angle ABC = \angle ACB$. Then triangle $ABC$ is congruent to triangle $ACB$ since the sides $BC$ and $CB$ are equal and the adjoining angles are equal.

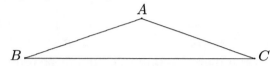

See also **explicit assertion**.

*Acknowledgments:* Atish Bagchi.

**permutation**    A permutation is defined in the literature in two different ways:

  a) A permutation of an $n$-element set is a sequence of length $n$ in which each element of the set appears once.
  b) A permutation of a set is a bijection from the set to itself.

Of course, the two definitions can be converted into each other, but psychologically they are rather different. Both definitions are given by [Kolman, Busby and Ross, 1996], pages 75 and 181. Compare the two different definitions of **equivalence relation**. *Citations:* (213), (392).

**Platonism**    Often used by mathematicians to refer to the attitude that **mathematical objects** exist in some manner analogous to the existence of physical objects.

All mathematicians, whether they regard themselves as Platonists or not, refer to **mathematical objects** using the same grammatical constructions as are used for references to physical objects. For example, one refers to "a continuous function" (indefinite reference) and "the sine function" (definite reference), in the way one refers to "a boy" and "the boss", not in the way one refers to nonmathematical abstract concepts

*Even empiricists must assert the existence of mathematical objects, exactly as that of physical objects, if they consistently apply their own criteria of existence.*

*–Kurt Gödel*

197

such as "truth" or "gravity" (no article). (This behavior is not limited to mathematical objects: "the orbit of the moon" for example.) Symbols are generally used in the same way as proper nouns.

See also **mathematical object** and Remark 2 under **symbol**. *Citations:* (2), (9), (20), (27), (70), (303), (410).

**plug into**   "Plug $a$ into $f$" means **evaluate** $f$ at $a$. Here, $f$ may be a **function** or an **expression**, and $a$ may be an expression.

***Example 1***   "If you plug $\pi$ into the sine function, you get 0."

***Remark 1***   Some find the use of the phrase "plug into" offensive. I judge this phrase to have low **status**. *Citations:* (231), (289), (401).

**plural**   Many authors form the plural of certain learned words using endings from the language from which the words originated. Students may get these wrong, and may sometimes meet with ridicule for doing so.

***(a) Plurals ending in a vowel***   Here are some of the common mathematical terms with vowel plurals.

| *singular* | *plural* |
|---|---|
| automaton | automata |
| polyhedron | polyhedra |
| focus | foci |
| locus | loci |
| radius | radii |
| formula | formulae |

Linguists have noted that such plurals seem to be processed differently from s-plurals ([Pinker and Prince, 1999]). In particular, when used as adjectives, most nouns appear in the singular, but vowel-plural nouns appear in the plural: Compare "automata theory" with "group theory".

The plurals that end in a (of Greek and Latin neuter nouns) are often not recognized as plurals and are therefore used as singulars. (This does not seem to happen with my students with the -i plurals and the -ae plurals.)

In the written literature, the -ae plural appears to be dying, but the -a and -i plurals are hanging on. The commonest -ae plural is "formulae"; other feminine Latin nouns such as "parabola" are usually used with the English plural. In the 1990–1995 issues of six American mathematics journals (American Journal of Mathematics, American Mathematical Monthly, Annals of Mathematics, Journal of the American Mathematical Society, Proceedings of the American Mathematical Society, Transactions of the American Mathematical Society), I found 829 occurrences of "formulas" and 260 occurrences of "formulae", in contrast with 17 occurrences of "parabolas" and and no occurrences of "parabolae". (There were only three occurrences of "parabolae" after 1918.) In contrast, there were 107 occurrences of "polyhedra" and only 14 of "polyhedrons".

> It is not unfair to say that some scholars insist on using foreign plurals as a form of one-upmanship. Students and young professors need to be aware of these plurals in their own self interest.
>
> It appears to me that ridicule and put-down for using standard English plurals instead of foreign plurals, and for mispronouncing foreign names, is much less common than it was thirty years ago. However, I am assured by students that it still happens.

### (b) Plurals in s with modified roots

| singular | plural |
|----------|--------|
| matrix   | matrices |
| simplex  | simplices |
| vertex   | vertices |

Students recognize these as plurals but produce new singulars for the words as **back formations**. For example, one hears "matricee" and "verticee" as the singular for "matrix" and "vertex". I have also heard "vertec".

The use of plurals in the **mathematical register** is discussed under **collective plural** and **distributive plural**.

*Acknowledgments:* Atish Bagchi, Eric Schedler.

**pointwise**   See coordinatewise.

**pointy bracket**   See delimiter.

**Polish notation**   **Polish notation** consists in using prefix notation without parentheses. This requires that all function names have a single arity, so that which symbols apply to which inputs is unambiguous.

*Example 1*   In Polish notation,
$$2 \sin x + \sin y$$
would be written
$$+ * 2 \sin x \sin y$$
with $*$ denoting multiplication.

    See also reverse Polish notation.

> Polish notation originated with the Polish logic school of the 1920's. In particular the phrases "Polish notation" and "reverse Polish notation" originated from that fact and were not intended as ethnic slurs.

*Terminology*   Some authors use the phrase "Polish notation" even if parentheses are used (they are always redundant but add intelligibility). Polish notation is occasionally called **left Polish notation**. *Citations:* (374), (290), (32).

**polymorphic**   See overloaded notation.

**pons asinorum**   The theorem in plane geometry that if a triangle has two equal angles then it has two equal sides has been called the **pons asinorum** (bridge of donkeys) because some students found its proof impossible to understand. A proof is discussed on page 197.

> I assume, like Coxeter [1980], that the name "bridge" comes from the fact that an isosceles triangle looks like a old arched stone footbridge, particularly if the triangle is drawn wider than it is tall as on page 197.

    The problem of proving the Pythagorean Theorem has also been called the pons asinorum.

*Difficulties* It would be worthwhile to find out what are the **concepts** or **proofs** that could be called a pons asinorum for modern undergraduate math majors. Some possibilities:

- The difference between **continuity** and uniform continuity eludes some students. This is probably because of a general problem they have with **order of quantifiers**.
- Students find it difficult to **reify** the equivalence classes of an **equivalence relation**. It is a standard tool in higher mathematics to take the classes of a partition and make them elements of a structure, points in a **space**, and so on, for example in the construction of quotient spaces or groups. Students may not understand that (for example in a quotient group) one must think of multiplying the cosets *themselves* instead of multiplying their elements.

**positive** In most (but not all) North American texts and university courses, the phrase "$x$ is positive" means $x > 0$. In a European setting it may mean $x \geq 0$. See **convention** and **private language**. This may have been an innovation by Bourbaki. Similar remarks may be made about **negative**.

**postcondition** A **postcondition** in a **definition** or statement of a theorem is a condition stated after the definition or theorem.

> "If $n$ is divisible by four then it is even. This holds for any integer $n$."

The second sentence is a postcondition. Another example is given under **where**. *Citations:* (236), (244), (254), (275), (343).

**postfix notation** Postfix notation consists in writing the name of the **function** after its **arguments**.

**Example 1**  The expression $x+y$ in postfix notation would be $(x, y)+$. *Citations:* (426), (139).

Most authors write functions of one **variable** in **prefix notation**, but some algebraists use postfix notation. The symbol "!" denoting the factorial function is normally written in postfix notation.

See also **Polish notation**, **prefix notation** and **rightists**.

**power**  The integer $5^3$ is a **power** of 5 with exponent 3. One also describes $5^3$ as "5 to the third power". I have seen students confused by this double usage. A statement such as "8 is a power of 2" may make the student think of $2^8$. *Citations:* (335), (347).

**precedence**  If $\Delta$ and $*$ are two **binary operators**, one says that $\Delta$ has higher **precedence** than $*$ if the **expression** $x\Delta y * z$ denotes $(x\Delta y) * z$ rather than $x\Delta(y * z)$.

**Example 1**  The expression $xy + z$ means $(xy) + z$, not $x(y + z)$, because in the **symbolic language**, multiplication has higher precedence than addition.

Unary operations (**functions** with one **input**) in mathematical writing typically have low precedence.

The **metaphor** behind the word "precedence" is that if one carries out a calculation of the expression, one must apply the operator with higher precedence before the other one. Thus in calculating $(x\Delta y) * z$ one calculates $u = x\Delta y$ and *then* $u * z$.

**Example 2**  One writes $\sin x$ but $\sin(x+y)$ because $\sin x+ y$ may be perceived as either ambiguous or as $(\sin x) + y$. As this example illustrates, in the traditional **symbolic language** the precedence relationship of some pairs of operations is not necessarily **well-defined**. *Citation:* (221).

See **delimiter** and **evaluation**.

**predicate symbol**  A symbol may be used in **mathematical logic** to denote an **assertion** containing **variables**. It is then called a **predicate symbol**, or just a **predicate**.

*Example 1*   The familiar binary relations $<$, $\leq$, $>$ and $\geq$ are predicate symbols. The expressions $3 < 5$, $3 \leq 5$, $3 > 5$ and $3 \geq 5$ are all assertions (although only the first two are *correct* assertions). *Citations:* (126), (287).

*Example 2*   Mathematics texts may use **mnemonic** predicate symbols. Thus a text might use a predicate symbol Sq, meaning "is a square", in assertions such as Sq(9), meaning "9 is a square." *Citation:* (278).

**prefix notation**   An **expression** is in **prefix notation** if the function symbols are written on the left of the **argument**.

*Example 1*   The expression $x + y$ written in prefix notation would be $+(x, y)$

*Remark 1*   In the traditional mathematical **symbolic language**, functions of one **variable** are used in prefix notation but a few, for example the symbols for the factorial and the greatest integer function, are used in other ways. Most **binary operations** denoted by special nonalphabetical symbols are written in **infix notation**, but those with alphabetical symbols are generally written in **prefix notation** and others such as an inner product may be written in **outfix notation**. *Citations:* (12), (426).

    See also **postfix notation**, **Polish notation**, **reverse Polish notation** and **outfix notation**.

**prescriptivist**   A **prescriptivist** is someone who gives rules for which forms and **syntax** are correct in English (or another language). Prescriptivists are those who say we should not use double negatives, split infinitives, and "ain't". Opposed to **descriptivist**.

    **Esilism** is a special form of prescriptivism.

**prime**   (The typographical symbol). The symbol "$'$" is pronounced "prime" or "dash". For example, $x'$ is pronounced "x prime" or "x dash".

$x''$ is pronounced "x double prime". The pronunciation "dash" is used mostly outside the USA.

**(a) Indicates a new variable** If $x$ is a variable of a certain type, then $x'$ and $x''$ are new variables of the same type.

**Example 1** "Let $S$ be a set and $S'$ and $S''$ subsets of $S$." *Citations:* (10), (122), (155), (180), (311), (325).

**(b) Derivative** If $f$ is a differentiable function, its derivative may be indicated by $f'$. See irregular syntax. *Citations:* (49), (147), (192), (362).

**(c) Other uses** The prime is sometimes used for functional operations other than the derivative, for example the Boolean complement or the derived set of a subset of a topological space. *Citation:* (68),

**private language** Sometimes an author or teacher will give a different definition to a term that has acquired a reasonably standard meaning. This may even be done without warning the reader or student that the definition is deviant. I would say that the person doing this is using a **private language**. Such an author has no sense of being in a community of scholars who expect to have a common vocabulary: to use recent slang, the author is on "another planet".

*Maybe this world is
another planet's hell.*
*—Aldous Huxley*

**Example 1** It has been standard usage in the research literature for fifty years to write $A \subset B$ to mean that $A$ is included as a subset in $B$, in particular allowing $A = B$. In recent years, authors of high school and lower-level college texts commonly write $A \subseteq B$ to mean that $A$ is included in $B$. *Citation:* (388). Some of these write $A \subset B$ to mean that $A$ is properly included in $B$ (citation (377)), thereby clashing with the usage in research literature. This was probably the result of formal analogy.

Using $A \subset B$ to mean $A$ is properly included in $B$ seems to be much less common that the usage of "$\subseteq$" and in my opinion should be deprecated.

*Acknowledgments:* Michael Barr, Eric Schedler.

**process**   See APOS and algorithm.

**program**   See algorithm.

**pronunciation**   Some students have told me that they find it necessary to be able to pronounce an **expression** that occurs in a **text**; if they can't, they can't read the text. One student brought this up with the common notation "$F : S \to T$".

This difficulty surprised me. Upon introspection, I discovered that when I was reading mathematics the inner voice in my head simply went silent at certain constructions, specifically in the case of "$F : S \to T$". Clearly some readers can't tolerate this.

See **context-sensitive**, **mental representation** and **names from other languages**.

**procept**   See object-process duality.

**proof**   A **proof** is a step by step argument intended to persuade other mathematicians of the correctness of an **assertion**. Proofs may be in **narrative style** or **labeled style**, discussed under those headings.

The individual phrases in a proof can be classified as follows:

*(i) Proof steps*   A proof will contain formal mathematical statements that contain calculations or follow from previous statements. We call these **proof steps**. They are assertions in the mathematical register, like theorems, but unlike theorems one must deduce from the **context** the hypotheses that make them true.

*Proof is an idol before which the mathematician tortures himself.*
*– Arthur Eddington*

*(ii)* **Reasons**  Statement of a previous result in this same proof that is the justification for either the next proof step or the one just stated.

*(iii)* **Summaries**  These state what must be proved, what we are just about to prove, or, part way through a proof, what is left to be proved or what has just been proved.

*(iv)* **Pointers**  **Pointers** give the location of pieces of the proof that are out of order, either elsewhere in the current proof or elsewhere in the **text** or in another text. References to another text are commonly called **citations**.

**Example 1**  I give a short narrative proof to illustrate these ideas. Define $m \mid n$ (read "$m$ divides $n$") to mean that there is an integer $q$ for which $n = qm$, and define $m \cong m' \pmod{k}$ to mean that $k \mid (m - m')$. Assume that I have already proved

*Lemma A*: If $m \mid n$ and $m \mid n'$ then $m \mid (n + n')$.

I will prove that if $m \cong m' \pmod{k}$ and $n \cong n' \pmod{k}$ then $m + n \cong m' + n'$ and $mn \cong m'n' \pmod{k}$. Proof:

| | |
|---|---|
| *Reason* | The **hypothesis** translates into the **assertion** |
| *Proof Step* | $k \mid m - m'$ and $k \mid n - n'$. |
| *Proof Step* | Now $(m + n) - (m' + n') = m - m' + n - n'$; |
| *Reason* | this is the sum of two numbers divisible by $k$, so |
| *Pointer* | by Lemma A |
| *Proof Step* | it is divisible by $k$. |
| *Proof Step* | Hence $m + n \cong m' + n' \pmod{k}$. |
| *Summary* | Now we must show that $mn \cong m'n' \pmod{k}$. |
| *Reason* | A little algebra shows that |
| *Proof Step* | $mn - m'n' = m(n - n') + n'(m - m')$. |
| *Pointer* | By Lemma A, |
| *Proof Step* | $k \mid (mn - m'n')$, |
| *Reason* | so by definition, |

*Proof Step*    $mn \cong m'n'$ (mod $k$).

This discussion is drawn from [Bagchi and Wells, 1998a]. [Hanna, 1992] discusses the role of proofs in mathematics (with lots of references to the literature) and issues for **mathematical education**. Other discussions of proof in mathematical education may be found in [Dreyfus, 1999], [Epp, 1998], [Nardi, 1998], [Tall, 1999], [Tall, 2002].

**_Proofs in mathematical logic_**  Mathematical logic also has a concept called **proof**: that is a **mathematical object** intended to **model** mathematicians' proofs. Proofs in mathematical logic may be called **formal proofs**, but that phrase is also used to indicate a particularly careful and detailed proof in the ordinary sense.

**proof by contradiction**  There are two somewhat different formats for **proof** that mathematicians refer to as proof by contradiction.

**(a) _Proof by deducing a false statement_**  To prove $P$, assume $P$ is false and deduce some **assertion** $Q$ that is known to be false. This is the form of one well-known proof that $\sqrt{2}$ is irrational; one assumes it is rational and then concludes by violating the known fact that every fraction can be reduced to lowest terms.

Authors, even writing for undergraduates, often give such a proof by contradiction without saying they are doing it. The format of such a proof would be:

a) Theorem: $P$.

b) Assume $P$ is false.

c) **Argument** that $R$ is true, where $R$ is some **statement** well known to be false. The argument that $R$ is true will assume that $P$ is false, possibly without saying so.

d) End of proof.

The student must recognize the pattern of proof by contradiction without being told that that is what it is.

*He hoped and prayed that there wasn't an afterlife. Then he realized there was a contradiction involved here and merely hoped that there wasn't an afterlife.*
　　　　*–Douglas Adams*

***Example 1*** I will prove the $\sqrt{2}$ is irrational using this format. I will use the fact that if $k^2$ is even then so is $k$. Assume that $\sqrt{2} = \frac{m}{n}$. Then $\frac{m^2}{n^2} = 2$, so $m^2 = 2n^2$. Therefore $m^2$ is even, so $m$ is even. Hence $m = 2k$ for some $k$. Therefore $2n^2 = 4k^2$, so $n^2 = 2k^2$. Hence $n$ is even. Thus any way of writing $\sqrt{2}$ as a fraction of **integers** has the result that both the numerator and denominator are even, which contradicts the fact that a fraction can always be reduced to lowest terms.

In the first sentence, I assume that $\sqrt{2}$ is rational without saying that we are aiming at a proof by contradiction. Furthermore, the assumption that $\sqrt{2}$ is rational is stated by **rewriting using the definition**, again without saying so. This sort of thing is very common in the literature.

See **pattern recognition**.

***(b) Proof by contrapositive*** A proof that a **conditional assertion** $P \Rightarrow Q$ is **true** may be explicitly called a proof by contradiction but will follow the format for a proof by **contrapositive**, omitting step (d) in the format given in Remark 1 under **contrapositive**.

*Citations:* (245). See also Example 7 under **let**.

*Reference:* [Krantz, 1997], page 68, discusses how to write proofs by contradiction.

*Acknowledgments:* Atish Bagchi, Eric Schedler.

**proof by instruction** A **proof by instruction** consists of instructions as to how to write a proof or how to modify a given proof to obtain another one. They come in several types.

***(i) Geometric instructions*** As an example, I could have worded the proof in Example 5 under **pattern recognition** this way: "Flip triangle $ABC$ around the bisector of side $BC$ and you must get the same triangle since a side and two adjoining angles are equal. Thus $AB = AC$."

***(ii) Algebraic instructions***   An example is the instruction in Example 1 under **look ahead** to divide the Pythagorean identity $a^2 + b^2 = c^2$ by $c^2$ to obtain the identity $\sin^2 \theta + \cos^2 \theta = 1$. The proof in Example 1 under **proof** has a miniproof by instruction in line 9. *Citations:* (13), (329).

***(iii) Directions for modifying a proof***   This is an instruction such as "interchange the role of $x$ and $y$ in the preceding proof and you get a proof of ... ". *Citation:* (132).

**proper**   A subset $T$ of a set $S$ is a **proper** subset if it is not $S$. This is also used with substructures of a structure (proper subgroup, and so on).

The word is also used to mean **nontrivial**; for example, a proper automorphism would be a non-identity automorphism. *Citations:* (256), (347).

**property**   A **property** that an instance of a class of **mathematical objects** may have determines a subclass of those objects.

***Example 1***   Being even is a property that integers may have. This property determines a subset of integers, namely the set of even integers.

One states that an object has a property using a form of "to **be**" and an adjective or a **noun phrase**.

***Example 2***   One can say

"4 is even."

or

"4 is an even integer."

*Citations:* (173), (249), (302), (423).

In some cases the property may also be given by a verb. See **vanish** for examples.

***Property as subclass*** Mathematical texts sometimes **define** a property as the class of **objects** having that property. Thus the property of being even *is* the set of even integers.

This has some consequences. For one thing, two different definitions can give the same property, so that a property is (perhaps unexpectedly) not a linguistic concept. For example, say an **integer** $n$ is "squeeven" if $n^2$ is even. Now, $n^2$ is even if and only if $n$ is even, so according to the definition of property as a subclass, squeeven is *the same property* as even, although its definition is different.

Another consequence is that there are more properties for elements of an **infinite** set than there are possible **names** for them.

In my experience, this point of view causes students quite a bit of difficulty. On the other hand, defining "property" in any language-oriented way appears to be complicated and difficult.

See also **relation**.

Some authors and editors object to using a property named after a person as a predicate adjective. Instead of saying "The space $S$ is Hausdorff" they would prefer "$S$ is a Hausdorff space."

**proposition** **Proposition** is used as another word for **theorem**. Some texts distinguish between propositions and theorems, reserving the word "theorem" for those that are considered especially important. This is the practice in [Epp, 1995], for example (see her discussion on page 129). *Citations:* (123), (253).

The word "proposition" is used in some texts to denote any **assertion** that is definitely true or definitely false. *Citations:* (123), (412).

**prototype** Commonly a human **concept** has **typical** members, called **prototypes** by Lakoff.

***Example 1*** For many people, a sparrow is a prototypical bird, and a penguin is not.

*The art of doing mathematics consists in finding that special case which contains all the germs of generality.*
*–David Hilbert*

***Example 2***   Students tend to have a prototype of the concept of limit of a **sequence** in which the entries in the sequence never take on the value of the limit. This is discussed under **limit**.

    The concept of "prototype" is subtle; these examples only hint at its depth. See also **example** and **radial concept**.

**provided that**   Used like if to give a definition.

***Example 1***   "The integer $n$ is **squarefree** provided that no square of a prime divides $n$." Also **providing that**. *Citations:* (90), (388).

    *Acknowledgments:* Atish Bagchi.

**put**   Used in **definitions**, mainly to define a **symbol**. **Set** is used in the same way.

***Example 1***   "Put $f(x) = x^2 \sin x$."

    *Citation:* (29).

**Q**   The symbol $\mathbb{Q}$ usually denotes the **set** of rational numbers. *Citation:* (104).

**quantifier**   In this text, a **quantifier** is either the **existential quantifier** or the **universal quantifier**. Linguists and logicians study other quantifiers not discussed here. See for example [Chierchia and McConnell-Ginet, 1990], Chapter 9, [Hintikka, 1996], and [Henkin, 1961].

**R**   The symbol $\mathbb{R}$ usually denotes the **set** of real numbers. Sometimes it is used for the rationals. *Citations:* (42), (303), (315).

**radial concept**   A **radial concept** or **radial category** is a **concept** with some central **prototypical** examples, and also some objects that deviate from the prototypical examples in various ways. Some deviant objects may be called by the basic name whereas others could not be (see Example 1 below).

Some members of a radial category deviate only slightly from the prototypes, others are highly metaphorical, and some contrast in some way with the prototypes. The members are not necessarily automatically generated from the prototypical examples; membership is to a considerable extent a matter of **convention**.

> Our mental representation of the world is to a great extent organized around radial categories. The practice of adding new deviant members to a radial category is common and largely unconscious.

**Example 1**  The concept of "mother" is a radial concept. Various members of the category among English-speakers include birth mother, adoptive mother, foster mother, earth mother, house mother, stepmother, grandmother, expectant mother and mother-in-law. Note that typically an adoptive mother is in our culture commonly referred to as a mother, but a house mother might not be called a mother. Aside from these there are words such as motherboard, mother lode and mother of pearl that seem to me to have a different status from those in the the first list (and they definitely would not be called mothers) but which some authors would classify as part of the category.

**Example 2**  Many phrases used by mathematicians are instances of radial categories. Consider

- **Incomplete proof** (compare expectant mother).
- **Multivalued function** (compare Triple Crown).
- **Left identity** (compare half-brother).

According to the very special way mathematical concepts are formulated, by **accumulation of attributes**, an incomplete proof is not a proof, a multivalued function is not a function, and a left identity may not be an identity. **Literalists** tend to object to such usages, but they are fighting a losing battle against a basic method built into the human brain for organizing our **mental representation** of the world.

*Citations:* (209), (177), (253).

*Acknowledgments:* The name "radial" and the mother examples come from [Lakoff, 1986]. Also thanks to Gerard Buskes.

**range**   Depending on the text, the **range** of a function can denote either the **codomain** or the **image**. The texts [Krantz, 1995] takes the first approach, and [Epp, 1995] and [Grassman and Tremblay, 1996] take the second approach.

**ratchet effect**   Once you acquire an insight, you may not be able to understand how someone else can't understand it. It becomes **obvious**, or **trivial** to prove. That is the **ratchet effect**.

This process probably involves synthesizing a new concept, as discussed by Dreyfus [1992], section 3.2. See also [Thurston, 1990].

***Remark 1***   It is distressingly common that a mathematician for whom a concept has become obvious because of the ratchet effect will then tell students that the concept is obvious or trivial. This is the phenomenon discussed in the sidebar under **trivial**. It is a major point made in [Kenschaft, 1997], page 30.

**real number**   **Real numbers** are associated have an elaborate **schema** that students are expected to absorb in part by **osmosis**. Some aspects of this schema which cause problems for students are listed here.

- **Integers** and rational numbers are real numbers (see the sidebar).
- A real number represents a point on the real line.
- A real number has a decimal expansion, but the **representation** is not bijective because of the infinite sequence of nines phenomenon.
- The decimal expansion is not itself the real number. See item (vii) under **behaviors**.
- Part of the decimal expansion of a real number approximates the number in a precise way. However, some numbers (for example, 1/3 and $\sqrt{2}$) are defined exactly even though no part of their decimal expansion gives the number exactly.

- The four basic arithmetic operations are defined for real numbers, although it is not obvious how to carry out the usual **algorithms** when the expansions of both numbers are infinite and nonrepeating.
- There is always another real number between any two distinct real numbers.

The word **real** is frequently used as a predicate adjective, as in "Let $x$ be real," meaning $x$ is a real number. I have heard students use the phrase "real number" to mean "genuine number", that is, not a variable. See also **space**. *Citations:* (85), (342).

Computer languages typically treat integers as if they were distinct from real numbers. In particular, many computer languages have the **convention** that the expression 2 denotes the integer and the expression 2.0 denotes the real number. I have known students who assumed that professors of mathematics were all familiar with this fact (probably nearly true in recent years) and that we all use notation this way (almost never true).

**recall**    Used before giving a **definition**, **theorem** or **proof**.

*Example 1*    "Recall that an integer is even if it is divisible by 2." The intent seems to be that the author expects that the reader already knows the meaning of the defined term, but just in case here is a reminder. See Remark 4 under **mathematical definition**. *Citations:* (137), (146).

**reconstructive generalization**    See generalization.

**reductionist**    A **reductionist** or **eliminativist** believes that all mathematical concepts should be reduced to as few concepts as possible, at least for purposes of foundations of mathematics. This is usually done by reducing everything to sets. Not all mathematicians (especially not category theorists) agree with this approach.

This becomes a problem only if the reductionists insist on thinking of **mathematical objects** in terms of their reductions, or on insisting that their reductions *are what they really are.* See **literalist**. Most mathematicians, even those who agree with reductionism for foundational purposes, are more relaxed than this.

*Acknowledgments:* The word "eliminativist" comes from [Lakoff and Núñez, 2000].

## redundant

*(a) Redundancy in discourse*  A given discourse is **redundant** if it contains words and expressions that could be omitted without changing the meaning. As another example, consider the sentence

"The counting function of primes

$$\pi(x) := \# \{p \leq x : p \text{ prime}\}$$

satisfies the formula $\pi(x) \sim x/(\log x)$."

The phrase "the counting function of primes" is redundant, since the definition just following that phrase *says* it is the counting function for primes. This example, adapted from [Bateman and Diamond, 1996], is in no way bad writing: the redundancy adds to the reader's understanding (for *this* reader, anyway).

Type labeling is another commonly occurring systematic form of redundancy.

*(b) Redundancy in definitions*  Redundancy occurs in definitions in a different sense from the type of verbal redundancy just discussed. In this case redundancy refers to including **properties** or constituent structures that can be deduced from the rest of the definition.

*Structure determines* underlying set   An apparent systematic redundancy in definitions of **mathematical structures** occurs throughout mathematics, in that giving the structure typically determines the **underlying set**, but the definition usually mentions the underlying set anyway. (Rudin [1966] point out this phenomenon on page 18.)

*Example 1*   A **semigroup** is a set $S$ together with an associative **binary operation** $\star$ **defined on** $S$.

215

If you say what $\star$ is explicitly, what $S$ is is forced – it is the set of first (or second) coordinates of the domain of $\star$.

Similarly, if you give a topology, the underlying set is simply the maximal element of the topology.

In practice, however, it is common to specify the set when defining an operation.

**Example 2** The **cyclic group of order three** is defined up to isomorphism as the group with underlying set $\{0, 1, 2\}$ and multiplication given by addition mod 3.

Addition mod 3 defines a binary operation on the set
$$\{0, 1, 2, 3, 4, 5\}$$
as well, so the mention of the underlying set is necessary.

The point of this example is that if you give the operation **extensionally**, the operation does indeed determine the underlying set, but in fact operations are usually given by a rule that may not determine the underlying set uniquely.

*Other examples* There are some other examples in which the definition is redundant and the redundancy cannot be described as a matter of convention. For example, in defining a **group** one usually requires an identity and that every element have a two-sided inverse; in fact, a left identity and left inverses with respect to the left identity are enough. In this case it is properties, rather than data, that are redundant. See radial concept.

> I have heard mathematicians say (but not seen in print) that an assertion purporting to be a mathematical definition is not a definition if it is redundant. This is a *very* unwise stance, since it can be an unsolvable problem to determine if a particular definition is redundant. Nevertheless, for reasons of efficiency in proof, irredundant definitions are certainly desirable.

*Acknowledgments:* Michael Barr.

**register** A **register** in linguistics is a choice of grammatical constructions, level of formality and other aspects of the language, suitable for use in a given social context. The **scientific register** is the distinctive register for writing and speaking about science. It is marked in particular by the use of complex nominal phrases connected by verbs that describe relations rather than actions. That register and the difficulties students have with it is discussed in detail in [Halliday and Martin, 1993]. In that book the scientific register is called "scientific English", but the remarks in chapters 3 and 5 make it clear that the authors regard scientific English as a register.

A distinctive subregister of the scientific register is used in mathematics, namely the **mathematical register.**

**reification**   See object-process duality.

**relation**   Texts frequently define a (binary) **relation** on a set $S$ to be a subset of the cartesian product $S \times S$. The relation is however *used* as a two-place **infixed predicate symbol.**

***Example 1***   On the set of **real numbers**, let $R$ be relation

$$\{(x, x + 1) \mid x \in \mathbb{R}\}$$

Then $R$ is a relation, so for example the statement $3R4$ is true but the statement $3R5$ is false. These statements use $R$ as a predicate symbol, although it has been defined as a set. This caused much **cognitive dissonance** among my students. *Citation:* (126). See also [Chin and Tall, 2001].

**representation**   Mathematicians and their students make use of both **external representations** and **internal representations** of mathematical objects. These phrases are used in the mathematical education literature. I take internal representations to be the same thing as **mental representations** or concept images.

***External representations***   An external representation of a phenomenon is a mathematical or symbolic system intended to allow one to deduce

*PICTURE, n. A representation in two dimensions of something wearisome in three.*

*–Ambrose Bierce*

217

**assertions** about the phenomenon. Certain aspects of the phenomenon being represented are identified with certain **mathematical objects**; thus a representation involves a type of **conceptual blend**.

This is related to and may for some purposes be regarded as the same as the concept of **model**. Among mathematicians, the word "representation" is more likely to be used when **mathematical objects** are the phenomena being represented as well as the objects doing the representing, and "model" is more often used when physical phenomena are being represented by mathematical objects. This distinction must be regarded as preliminary and rough; it is not based on **citations**.

Logicians use "model" in a technical sense, roughly a **mathematical object** that fulfils the requirements of a theory. The "theory" in this case is itself a mathematical object, and the model is a type of homomorphism or functor that preserves logical reasoning.

***Example 1***   The expressions $10\frac{1}{2}$, 10.5 and $\frac{21}{2}$ are three *different* representations of the *same* mathematical object. See **value** and item (vii) under **behaviors**.

***Example 2***   Some of the ways in which one may represent **functions** are: as sets of ordered pairs, as **algorithms**, as maps (in the everyday sense) or other pictures, and as black boxes with input and output. Some of these representations are **mathematical objects** and others are mental representations. Other examples occur under **model**.

It may be seen from these examples that the internal and external representations of an idea are not sharply distinguished from one another. In particular, the internal representation will in general involve the symbolism and terminology of the external representation, as well as nonverbal and nonsymbolic images and relationships.

The book [Janvier, 1987] is a primary source of information about representations. [Thompson, 1994] discusses confusions in the concept of representation on pages 39ff. The need to keep in mind multiple representations is part of the discussion in [Thompson and Sfard, 1998]. See also [Vinner and Dreyfus, 1989].

***Remark 1***  Of course, "representation" is also a mathematical word with various definitions in different disciplines. These definitions are generally **abstractions** of the concept of representation discussed here.

**respectively**  Used to indicate term-by-term **coreference** between two lists of objects. Rarely used with lists with more than two entries.

***Example 1***  "The smallest prime divisors of 9 and 10 are 3 and 2, respectively." *Citations:* (78), (223), (246), (355).

See also **comma**, as well as **citation** (313).

**result**  The **value** produced by a **function** at a given input may be called the **result** of the function at that input. *Citation:* (289).

The word is also used to denote a mathematical fact that has been proved. *Citations:* (48), (202), (230).

**reverse Polish notation**  A form of **postfix notation** that is used without parentheses. This requires that the **arity** of all the symbols used be fixed.

***Example 1***  In reverse Polish notation,

$$2 \sin x + \sin y$$

would be written

$$x \sin 2 * y \sin +$$

with $*$ denoting multiplication. *Citation:* (374).

See **Polish notation**. Reverse Polish notation is sometimes called **right Polish notation**.

> Reverse Polish notation is used by some Hewlett-Packard calculators and by the computer languages Forth and Postscript. It has come into prominence because **expressions** in a reverse Polish language are already in the form that makes it easy to design an interpreter or compiler to process them.

**revise**  In the United States, to "revise" a document means to change it, hopefully improving it in the process. Speakers influenced by British

English use "revise" to mean "review"; in particular, students may talk about revising for an upcoming test. In this case there is no implication that anything will (or will not) be changed.

**Remark 1**   This entry has nothing directly to do with mathematics or the **mathematical register**, but I have several times witnessed the confusion it can cause in academic circles and so thought it worth including here. *Citation:* (339)

**rewrite using definitions**   One of the secrets of passing a first course in abstract mathematics that teaches **proofs** (first algebra course, first discrete math course, advanced calculus, and so on) is to take every statement to be proved and first rewrite it using the **definitions** of the terms in the statement. It is remarkably difficult to convince students to try this.

*Always substitute mentally the definitions in place of the defined.*
                    *–Blaise Pascal*

**Example 1**   A **relation** $\alpha$ is defined on the set of real numbers by

$$x \ \alpha \ y \ \text{if and only if} \ x < y - 1$$

Prove that $\alpha$ is transitive.  Proof: Rewrite the definition of transitive: We must show that if $x \ \alpha \ y$ and $y \ \alpha \ z$ then $x \ \alpha \ z$. Rewrite using the definition of $\alpha$: This means we must show that if $x < y - 1$ and $y < z - 1$ then $x < z - 1$. The hypotheses show that $x < y - 1 < (z - 1) - 1 < z - 1$ as required.

This technique is useful for finding **counterexamples**, as well. Try it when $\alpha$ is defined by

$$x \ \alpha \ y \ \text{if and only if} \ x < y + 1$$

Another example is given under **trivial**. See also **unwind**.
*Acknowledgments:* Eric Schedler.

**rightists**   Occasionally naive young authors start using **postfix notation** for functions so that functional **composition** can be read in its natural order. They are **rightists** (my name). Thus they will write $xf$ instead of $f(x)$ or $fx$, hence $xfg$ instead of $g(f(x))$, allowing them to write the composite as $f \circ g$ instead of $g \circ f$. Obviously, they think, *everyone* ought to do this. Then they get complaints from people who find their papers hard or impossible to read, and they revert to the usual **prefix notation**. I was one of those naifs in the 1970's.

Postfix notation has in fact caught on in some fields, particularly some branches of abstract algebra.

Some French authors stick with prefix notation but then express functions in the **straight arrow notation** so that the horizontal arrows go to the left: thus $f : B \longleftarrow A$ or more commonly

$$B \xleftarrow{\quad f \quad} A$$

instead of $f : A \to B$. They draw commutative diagrams in that direction, too. Then $g \circ f$ is pictured as

$$C \xleftarrow{\quad g \quad} B \xleftarrow{\quad f \quad} A$$

instead of

$$A \xrightarrow{\quad f \quad} B \xrightarrow{\quad g \quad} C$$

However, most mathematicians who use straight arrow notation stick with the latter form even though they write the composite as $g \circ f$.

See also **private language**. *Reference:* [Osofsky, 1994].

**root**    A **root** of an equation $f(x) = 0$ is a value $c$ for which $f(c) = 0$. This value $c$ is also called a root or a **zero** of the function $f$.

***Remark 1***   Some hold it to be incorrect to refer $c$ as a "root of $f$" instead of "zero of $f$". The practice is nevertheless quite widespread, particularly when the function is a polynomial. *Citations:* (63), (328).

***Remark 2***   "Root" is of course used with a different but related meaning in phrases such as "square root", "$n$th root", and so on.

*Acknowledgments:* Gary Tee.

**round parentheses**   See delimiter.

**sanity check**   A simple test to check if something you have formulated makes sense.

***Example 1***   If you write down $6s = p$ for the **student-professor problem** and check your work by **plugging in** $s = 12$, $p = 2$, you immediately discover your error.

**satisfy**   A mathematical structure **satisfies** an **assertion** that contains **variables** if the assertion makes a meaningful statement about the structure that becomes **true** for every possible **instantiation** of the variables.

***Example 1***   "Every **group** satisfies the statement $\forall x \exists y (xyx = x)$."
*Citations:* (9), (141), (234), (384).

**say**

*(a) **To signal a definition***   **Say** may be used to signal that a **definition** is being given.

***Example 1***   "We say that an integer $n$ is **even** if $n$ is **divisible** by 2."
Variation:

  "An integer $n$ is said to be even if it is divisible by 2."

  *Citations:* (302), (357), (398).

*(b) **To introduce notation***   The word "say" is also used to introduce **notation**, especially to give a working name to a **variable object** used in a **universal generalization**.

***Example 2***   Let $f(x)$ be a polynomial with complex coefficients, say
$$f(x) = a_0 + a_1 x + \ldots + a_n x^n$$

One could then prove, for example, that $a_{n-1}$ is the sum of the roots of $f$ (counting multiplicity), a result that then holds for any complex polynomial.

*Citations:* (313), (283).

**Remark 1** Note that the syntax used with "say" for definitions is different from that for introducing notation.

*Acknowledgments:* Atish Bagchi, Eric Schedler.

## schema  See APOS.

## scope
The scope of an assumption and the scope of a local identifier are discussed in those entries.

In a symbolic expression, a variable is within the **scope** of an operator if its meaning or use is affected by the operator. I will discuss the use of this word here only for operators that **bind** variables.

**Example 1** In the expression

$$\int_a^b x^2 \, dx$$

the variable $x$ is bound by the integral operator.

**Example 2** In the expression

$$\int_a^b (x + y)^2 \, dx$$

the $x$ is bound but not the $y$, so that one would expect the value to be in terms of $a$, $b$ and $y$, but not $x$.

See bound variable.

**Remark 1** A mathematical definition of the scope of an operator, like that of bound variable, requires a formal recursive definition of "symbolic expression". The definition given in this entry is a dictionary definition. This is discussed in more detail in Remark 2 under free variable.

## self-monitoring
Self-monitoring is the activity a student engages in when she notices that some practice she uses in solving problems is

counterproductive (or is helpful) and modifies her behavior accordingly. It is discussed in [Resnick, 1987], [Schoenfeld, 1987b], and [Wells, 1995].

**semantic contamination**    The connotations or implicit **metaphors** suggested by a word or phrase that has been given a mathematical **definition** sometimes create an expectation in the reader that the word or phrase has a certain meaning, different from the correct meaning given by the **definition**. This is **semantic contamination**. It is a form of **cognitive dissonance**. In this case the two modes of learning in the definition of cognitive dissonance are learning the meaning from the definition and learning the meaning implicitly from connotations of the word used (which is a common mode of learning outside mathematics.) A mathematics student may suppress the information given by the definition (or by part of it) and rely only on the connotations.

*Example 1*    The word **series** conveys to some students the concept that is actually denoted in mathematics by the word **sequence**.

Other examples of semantic contamination are given under **conditional assertion**, **continuum hypothesis**, **contrapositive**, **formula**, **inequality** and **only if**.

*Reference:* [Hersh, 1997a] gives many examples of disparities between the ordinary meaning and the mathematical meaning of mathematical words. Any of them could be the source of semantic contamination.

*Terminology*    The name "semantic contamination" is due to Pimm [1987], page 88.

**semantics**    A **semantics** is a method of determining the meaning of an **expression** in a natural or artificial language or in a system of **notation**.

*(a) Semantics of symbolic expressions*    Symbolic expressions in the **mathematical register** have both **intensional** (note the spelling) and **ex-**

**tensional** semantics. Speaking *very* roughly, the intensional semantics carries information concerning how its meaning is constructed or calculated; the extensional semantics is merely the **object(s)** denoted by the expression.

**Example 1** The intensional interpretation of
$$\frac{3+5}{2}$$
in the **mathematical register** is something like:

    "The result of adding 5 and 3 and dividing the result by 2."

The extensional interpretation of that fraction is 4.

    There is more to this: see item (i) under "How one thinks of functions" in the entry on **function** and the examples under **equation**.

*(b) **How we handle** object-process duality* Mathematical discourse *routinely* avoids making any distinction between these three objects:

- An **expression**,
- An **intensional** interpretation of the expression,
- The **extensional** interpretation of the expression (that is the **mathematical object** it denotes).

A particular use of an expression may have any of these roles.

Many computer scientists use the word "semantics" to mean **interpretation**. In this book, a semantics is a *method* of interpretation, not a particular interpretation. In this connection it is commonly used as a singular noun. Semantics is also used to denote the study of meaning in a general sense.

**Example 2** "Both terms in the left side of the equation $10m + 15n = k$ are divisible by 5, so $k$ is divisible by 5." This sentence refers to the expression $10m + 15n = k$ itself.

**Example 3** "If $x + 3 = 2$, then $3 + x = 2$." Here the sentence deals with the intensional interpretation of the expressions.

***Example 4***   "Since $x > 2$, it follows that the **expression** $x^2 - 1$ is not negative." This means that the expression $x^2 - 1$ does not have a negative *value*; thus the sentence refers to the extensional interpretation of the expression.

This is how we handle **object-process duality** smoothly (see [Gray and Tall, 1994]), and to succeed in mathematics a students must become fluent in doing this, with very little explicit notice being given to the phenomenon. *Citations:* (8), (376), (387).

Sometimes, even a typographical entity is used to refer to a **mathematical object**.

***Example 5***   "Since $x > 2$, the **brackets** in the expression $(x^2 - 1) + (x^3 - 1)$ are both nonnegative."

***(c) Semantics in mathematical logic***   **Mathematical logic** typically constructs an **interpretation** of a **text** in some formal language. For example, an interpretation of the **symbolic assertion** $x + 2 = 7$ might take the universe of the interpretation to be the set of integers, and could interpret $x$ as 2. A familiar semantics for algebraic expressions causes it to be interpreted as the assertion that $4 = 7$, and under the usual method of determining truth for that assignment, this **statement** is "invalid" in that interpretation. If $x$ is interpreted as 5 then the symbolic assertion is valid for that interpretation. One also says that 5 **satisfies** the assertion but 2 does not.

***Difficulties***   The semantics of natural languages is currently the object of intensive study by linguists. Good starting places to find out about this are [Chierchia and McConnell-Ginet, 1990] and [Partee, 1996]. Some of what semanticists have learned sheds light on students' misunderstandings: see for example the related discussions of **definite article**, **indefinite article**, **universal quantifier** and **existential quantifier**.

*Citation:* (336).

**sentence**   In this book, the word **sentence** refers to a sentence in the English language. The word is also used in mathematical logic for a symbolic expression that denotes an **assertion**.

**sequence**   A infinite **sequence** of elements of a set $S$ is typically referred to in one of several ways:

- A sequence $s_1, s_2, \ldots$ of elements of $S$ [or "elements in $S$"].
- A sequence $(s_1, s_2, \ldots)$ of elements of $S$ (**Angle brackets** or even **braces** are sometimes used.)
- A sequence $(s_n)$ of elements of $S$.

The notation for finite sequence has similar variations.

The elements $s_i$ are referred to as **entries** or as **elements** of the sequence.

The starting point may vary, for example a sequence $s_0, s_1, \ldots$. It is easy to get confused by the meaning of the phrase "The $k$th entry of the sequence" if the sequence starts with some entry other than $s_1$.

See also **semantic contamination** and **subscript**. *Citations:* (28), (91), (348).

## set

### 1. Verb
Use in **definitions**, usually to define a **symbol**.

***Example 1***   "Set $f(t) = 3t^2$." **Put** is used in the same way. *Citation:* (11).

### 2. Noun
In abstract mathematics courses one may be tempted to "define" set, only to quail at the prospect of presenting Zermelo-Fränkel set theory. This may result in a total cop-out accompanied by mutterings about collections of things. One way out is to give a **specification** for sets. Two crucial properties of sets that students need to know are

a) A set is not the same thing as its elements.

b) A set is determined completely by what its elements are.

Most facts about sets as used in undergraduate mathematics courses are made reasonable by knowing these two facts. See also **element**, **empty set** and **setbuilder notation**. *References:* [Wells, 1995], [Wells, 1997].

***Difficulties***  In advanced mathematics course structures such as quotient **groups** are built on sets whose elements are sets; this requires **reifying** the sets involved. See [Lakoff and Núñez, 1997].

Students sometimes express discomfort when faced with sets that seem too **arbitrary**. See **yes it's weird**.

**setbuilder notation**  The expression $\{x \mid P(x)\}$ defines a **set**. Its elements are exactly those $x$ for which the **condition** $P(x)$ is **true**. (The **type** of $x$ is often deduced from the context.) This is called **setbuilder notation** (a low-**status** name) or **set comprehension** (a higher status but confusing name). The condition $P$ is called the **defining condition**. Setbuilder notation is a form of **structural notation**.

***Difficulties***  The basic rule of inference for setbuilder notation is that $P(a)$ is **true** *if and only if* $a \in \{x \mid P(x)\}$. This means in particular that if $P(a)$ then $a \in \{x \mid P(x)\}$, and if not $P(a)$, then $a \notin \{x \mid P(x)\}$. Students may fail to make use of the latter fact. This may be related to the phenomenon described under **collective coreference**.

***Variations***  A colon is used by some authors instead of a vertical line.

One may put an **expression** before the vertical line. This can be misleading.

***Example 1***  The set $\{x^2 \mid x \in \mathbb{R}$ and $x \neq 3\}$ *does* contain 9, because $9 = (-3)^2$.

Gries and Schneider [1993], Chapter 11, give examples that show that putting an expression before the vertical line can be ambiguous.

(Example 1 above is *not* ambiguous.) They introduce a more elaborate notation that eliminates the ambiguity. *Citation:* (139), and (for the colon variation) (66).

infinite 130
proof 205

**show**    To prove (see **proof**). Some scientists and possibly some high school teachers use "show" in a meaning that is something like "provide evidence for" or "illustrate". It appears to me that the collegiate level usage is that "show" is nearly always synonymous with "prove". See [Maurer, 1991], page 15. *Citation:* (119).

    One colleague has suggested that mathematicians use "show" when the proof has a strong intuitive component. This seems to fit with what I have observed as well.

**sign**    The word **sign** is used to refer to the symbols "+" (the **plus sign**) and "−" (the **minus sign**). It is also sometimes used for other symbols, for example "the integral sign". The word "sign" is also used to refer to the question of whether an expression represents a numerical quantity that is positive or negative.

**Example 1**   Let $f(x) = x^2$. Then for negative $x$, $f(x)$ and $f'(x)$ are opposite in sign. *Citations:* (1), (416).

**snow**    Professors (and other math students) sometimes try to intimidate the students by confronting them with unbelievable or difficult to understand assertions without preparing the ground, in order to make them realize just how wonderfully knowledgeable the professor is and what worms the students are. If the professor succeeds in making the student feel this way, I will say he has **snowed** the student.

    Notions of **infinite cardinality** are a favorite tool for such putdowns. Thus it is a scam to try to startle or mystify students with statements such as "There are just as many even integers as integers!" The would-be snower is taking advantage of the mathematician's use of "same number

*The world is governed more by appearances than realities, so that it is fully as necessary to seem to know something as to know it.*

    *–Daniel Webster*

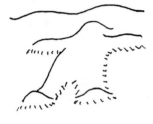

A THOROUGHLY
SNOWED STUDENT.

*The teacher pretend-*
*ed that algebra was a*
*perfectly natural affair,*
*to be taken for grant-*
*ed, whereas I didn't*
*even know what num-*
*bers were. Mathematics*
*classes became sheer ter-*
*ror and torture to me.*
*I was so intimidated*
*by my incomprehen-*
*sion that I did not dare*
*to ask any questions.*
                *–Carl Jung*

of elements" as a **metaphor** for sets in bijective correspondence, which in the case of infinite sets has properties at odds with the familiar properties of the idea for finite sets. (See [Lakoff and Núñez, 2000], pages 142–144.)

That scam is like asking a student "Please bring me that stick over there on the other blackboard" without mentioning the fact that you have decided to call a piece of chalk a "stick". It is true that there is *some* analogy between a piece of chalk and a stick (more than, say, between a piece of chalk and an elf), but I would expect the student to look confusedly for a long narrow thing made out of wood, not immediately guessing that you meant the piece of chalk.

**Remark 1** The successful student learns to resist being snowed. *Of course* the student does not know everything about mathematics. Neither does the professor. There are *always* things you don't know, and the more skillful would-be snowers manipulate the conversation so that they can talk about something they do know and their listener may not.

Unfortunately, I have known students who are what might be called **co-snowers.** They are all too ready to be humiliated by how little they know when someone however innocently refers to something they don't know about. Many of them drop out of mathematics because of this.

It is important to discuss this snow/co-snow phenomenon openly with students in beginning abstract math courses, where the problem is particularly bad (it seems to me). Airing the matter will surely give some of them heart to persevere.

**Remark 2** This use of the word "snow" is obsolescent slang from (I think) the sixties.

*Acknowledgments:* Eric Schedler.

**solution** A **solution** of an **equation** containing **variables** is a list of **instantiations** of all the variables that make the equation true.

**Example 1**   One of the two solutions of $x^2 = 2$ is $\sqrt{2}$. That is because if you substitute $\sqrt{2}$ into the equation you get the true **assertion** $(\sqrt{2})^2 = 2$.

**Example 2**   Every complex solution of the equation $x^2 = y^2$ is of the form $x = z$, $y = \pm z$. *Citations:* (49), (346).

**some**   The word **some** is used in the **mathematical register** to indicate the existential quantifier. Some examples are given under **existential quantifier**.

**space**   The naive concept of space based on our own physical experience is the primary **grounding metaphor** for some of the most important **structures** studied by mathematicians, including **vector** space, topological space, Banach space, and so on. The perception that **functions** can be regarded as points in a space (a function space) has been extraordinarily fruitful.

**Difficulties**   Most modern approaches to defining a certain kind of space as a mathematical structure begins with the concept that the space is a set of points with associated structure. This doesn't fit the naive picture of a space. The idea that points, which have zero size, can make up a space with *extent*, is completely counterintuitive. This shows up when students imagine, for example, that a real number has a "next" real number sitting right beside it (see Example 6 under **metaphor**). Students also have difficulty with envisioning the boundary (of zero width) of a subspace.

Intuitively, a space ought to be *a chunk with parts*, not *a collection of points*. The points ought to be *hard to see*, not the first thing you start with in the definition. This point of view has been developed in sheaf theory and in category theory. See [Lawvere and Schanuel, 1997] and [Lawvere and Rosebrugh, 2003].

*Points*
*Have no parts or joints*
*How then can they*
     *combine*
*To form a line?*
                    *–J.A. Lindon*

**specification** A **specification** of a mathematical concept describes the way the concept is used in sufficient detail for the purposes of a particular course or text, but does not give a mathematical **definition**. Specifications are particularly desirable in courses for students beginning abstract mathematics for concepts such as set, function and "ordered pair" where the standard definitions are either difficult or introduce irrelevant detail. Examples may be found under **set** and **function**.

**Remark 1** On pages 48ff of [Rota, 1996] the distinction is made between "description" and "**definition**" in mathematics. As an example of a description which is not a definition, he mentions D. C. Spencer's characterization of a tensor as "an object that transforms according to the following rules". That sounds mighty like a specification to me.

**Remark 2** Definitions in category theory, for example of "product", are often simply precise specifications. That is exemplified by the fact that a product of sets in the categorical sense is not a uniquely defined set in the way it appears to be in the classical definition as a set of ordered pairs. Category theory has made the practice of specification into a precise and dependable tool.

*References:* [Wells, 1995] and [Bagchi and Wells, 1998b].

**split definition** A definition by cases.

**square bracket** Square brackets are the delimiters []. They are occasionally used as **bare delimiters** and to enclose matrices, and may be used instead of parentheses to enclose the **argument** to a function in an expression of its **value** (as in $f[x]$ instead of $f(x)$). They are also used as **outfix notation** with other special meanings, for example to denote closed intervals. See **bracket**. *Citations:* (103), (115), (131), (189), (275), (381), (402).

**squiggle** See tilde.

**standard interpretation**  The **standard interpretation** of a mathematical discourse is the meaning a mathematician competent in a given field will understand from a discourse belonging to that field. (One aspect of being "competent", of course, is familiarity with the standard interpretation!)

I will state two theses about the standard interpretation here and make some comments.

> *There are no facts,*
> *only interpretations.*
> *–Friedrich Nietzsche*

**(a) First thesis**  There is such a thing as the standard interpretation and it is a proper subject for study in linguistics.

My evidence for this is that for most **mathematical discourse**, most mathematicians in the appropriate field who read it will agree on its meaning (and will mark students' papers wrong if they have a nonstandard interpretation). Furthermore, rules for how the interpretation is carried out can be apparently formulated for much of the symbolic part (see the discussion of Mathematica® under **symbolic language**), and some of the structure of the expressions that communicate **mathematical reasoning** has been the subject of intensive study by semanticists; for example, see [Chierchia and McConnell-Ginet, 1990] and [Kamp and Reyle, 1993].

> My claim that most of the time mathematicians agree on the meaning of what they read must be understood in the way that the claims of physics are understood. If an experiment disagrees with an established law, the experimenter can often discover a flaw that explains the disagreement. If mathematicians disagree about the meaning of a **text**, they often discover a flaw as well: one of them had the wrong definition of a word, they come to agree that the text is genuinely ambiguous, or the author tells them about a typo . . .

**(b) Second thesis**  One of the major tasks of an instructor in mathematics is to show a student how to extract correctly the standard interpretation of a piece of text.

This thesis is based on my own experience. I have always been sensitive to language-based misunderstandings, and not just in mathematics. I have kept records of such misunderstandings and learned some basic ideas of linguistics as a result of my curiosity about them. It appears to me from my teaching experience that language based misunderstandings cause problems in learning mathematics at the post-calculus level.

> *It is by universal misunderstanding that all agree. For if people understood each other, they would never agree.*
> *–Charles Baudelaire*

233

Many mathematical educators seem to advocate the point of view that the student's interpretation, however nonstandard, is just as valid as the mathematician's. There is much merit in not ridiculing the validity of the student's "misuse" of standard terminology, for example, those described under **conditional assertion** and **excluding special cases**, particularly when the misuse is customary linguistic behavior outside of mathematics. Nevertheless, it is vital that the student be told something like this: "Your usage may be perfectly natural in everyday discourse, but it is not the way mathematicians talk and so to be understood, *you must speak the standard dialect.*"

There is more about this in the entry **translation problem**.

**status**    I have had a few experiences that lead me to believe that some phrases in the **mathematical register** are "in" (have **high status**) and others are "out" (**low status**).

*Example 1*    To some mathematicians, "dummy variable" may sound high-schoolish and low status; it is much more refined to say "**bound variable**".

*Example 2*    The phrase "setbuilder notation" may have lower **status** than "set comprehension".

*Remark 1*    Variations in status no doubt differ in different mathematical disciplines.

*Remark 2*    I believe that in both examples just given, the low status word is much more likely to be understood by high school and beginning college students in the USA.

See also **plug into**.

A reviewer of a book I wrote said, " ... and he even referred to 'setbuilder notation' ... " without any further explanation as to why that was a bad thing. I was mightily puzzled by that remark until it occurred to me that status might be involved.

**straight arrow notation**    The notation $f : S \rightarrow T$ means that $f$ is a **function** with **domain** $S$ and **codomain** $T$. It is read "$f$ is a function from $S$ to $T$" if it is an independent clause and "$f$ from $S$ to $T$" if it is **parenthetic**. Compare **barred arrow notation**. See **rightists**.

*Citations:* (90), (403).

**structure**   See mathematical structure.

**structural notation**   Structural notation for a **mathematical object** is a **symbolic expression** that, in the given context, describes the (possibly variable) mathematical object unambiguously *without providing a* **symbol** *for it.* Also called **anonymous notation**.

**Example 1**   The **expression** $\{1, 2, 4\}$ is structural notation for the unique set that contains the elements 1, 2 and 4 and no other elements.

**Example 2**   The expression

$$\begin{pmatrix} a^2 & ab \\ -ab & b^2 \end{pmatrix}$$

is structural notation for a certain matrix with parameters $a$ and $b$.

**Example 3**   Setbuilder notation is a type of systematic structural notation. So are **barred arrow notation** and **lambda notation** for functions.

**subscript**   A string of **characters** is a **subscript** to a character if the string is placed immediately after the character and below the base line of the text. (But see Remark 1.)

**Example 1**   In the expression $x_{23}$, the string 23 is a subscript to $x$.

Subscripts are normally used for indexing.

**Example 2**   The tuple $a = (3, 1, 5)$ is determined by the fact that $a_1 = 3$, $a_2 = 1$, and $a_3 = 5$.

**Example 3**   The Fibonacci sequence $f_0, f_1, \ldots$ is defined by $f_0 = 0$, $f_1 = 1$, and $f_i = f_{i-1} + f_{i-2}$ for $i > 1$. (Some authors define $f_0 = 1$.) *Citations:* (28).

Subscripts may also be used to denote partial derivatives.

**Example 4**   If $F(x, y) = x^2 y^3$ then $F_x = 2xy^3$.

***Difficulties***  The tuple in Example 2 can be seen as a **function** on the set $\{1, 2, 3\}$ ("a tuple is a function on its index set"), and the Fibonacci sequence can be seen as a function on the nonnegative integers. The $i$th entry of Fibonacci sequence could thus be written indifferently as $f_i$ or $f(i)$. This fact is familiar to working mathematicians, but in a classroom where the Fibonacci function is denoted by $f_i$ a remark such as

"The Fibonacci sequence is an **increasing** function of $i$."

can cause considerable confusion to beginners.

***Remark 1***  Occasionally, as for example in dealing with tensors, a string is used as a **left subscript** by placing it immediately *before* the character and below the base line of the text.

***Example 5***  Let $_ka_i$ denote the $i$th coefficient of the $k$th polynomial $_kP$. *Citation:* (201).

*Acknowledgments:* Gary Tee.

**substitution**  To **substitute** an **expression** $e$ for a **variable** $x$ that occurs in an expression $t$ is to replace every occurrence of $x$ by $e$ (in a sophisticated way – see below). The expression resulting from the substitution has a possibly different **denotation** which can generally be determined from the **syntax**.

***Example 1***  Let $e$ be $x + y$ and $t$ be $2u$. Then substituting $t$ for $x$ in $e$ yields $2u + y$. *Citations:* (8), (286), (329).

***(a) Syntax of substitution***  The act of substituting may require insertion of **parentheses** and other adjustments to the expression containing the variable. In general, substituting is not a mechanical act, but requires understanding the **syntax** of the expression.

***Example 2***  Substituting $2u$ for $x$ in $x^2 + 2x + y$ gives $(2u)^2 + 2(2u) + y$; note the changes that have to be made from a straight textual substitution.

***Example 3***   Substituting 4 for $x$ in the expression $3x$ results in 12, not 34 (!).

***Example 4***   Suppose $f(x,y) = x^2 - y^2$. What is $f(y,x)$? What is $f(x,x)$? Many students have trouble with this kind of question.

See also **pattern recognition**.

     ***Uniform substitution***   Although a variate identifier denotes a **variable object**, or, if you wish, can be **instantiated** as any one of (usually) many possibilities, *all uses of a particular identifier refer to the same object.*

***Example 5***   "Let $f : S \to T$ be a surjective **function** between finite sets. Then $S$ has at least as many elements as $T$" This **assertion** is a **theorem**, which means that whenever you can **find** a function $f : A \to B$ with $A$ and $B$ finite sets, and you **substitute** $A$ for $S$ and $B$ for $T$, then the statement must be true.

     Now, $S$ (as well as $T$) occurs twice in the statement. The claim that the assertion is a theorem is based on substituting the *same* set $A$ for both occurrences of $S$, and similarly for $T$. Otherwise you could easily make the statement false.

> The only place in mathematics that I know of in which substitution of different objects is allowed for different occurrences of the same symbol is in the common notation for context-free grammars, as for example in [Hopcroft and Ullman, 1979].

***(b) Semantics of substitution***   A fundamental fact about the **syntax** and **semantics** of all mathematical **expressions** (as far as I know) is that *substitution commutes with* **evaluation**. This means that if you replace a subexpression by its value the value of the containing expression remains the same. For example, if you **instantiate** the **variable** $x$ in the expression $3x + y$ with 4 and replace the subexpression $3x$ by its value 12, you get the expression $12 + y$, which must have the same value as $3x + y$ as long as $x$ has the value 4. This is a basic fact about manipulating mathematical expressions.

     *Acknowledgments:* Michael Barr.

**subtract**   See minus.

**such that**   For a predicate $P$, a phrase of the form "$c$ such that $P(c)$" means that $P(c)$ holds.

***Example 1***   "Let $n$ be an integer such that $n > 2$." means that in the following **assertions** that refer to $n$, one can assume that $n > 2$.

***Remark 1***   At the blackboard, "such that" may be abbreviated "s.t." or denoted by "$\ni$" or sometimes "$:$". For the sentence in Example 1, one could write "Let $n$ be an integer s.t. $n > 2$" or "Let $n$ be an integer $\ni n > 2$." This is rarely done in print.

***Remark 2***   In pronouncing $\exists x P(x)$ the phrase "such that" is usually inserted. This is not done for the **universal quantifier**.

***Example 2***   "$\exists x(x > 0)$" is pronounced "There is an $x$ such that $x$ is greater than 0", but "$\forall x(x > 0)$" is pronounced "For all $x$, $x$ is greater than 0". *Citations:* (10), (24), (162).

> Yes, I know that "$\forall x(x > 0)$" is false.

   *Acknowledgments:* Susanna Epp and Kevin Weidenbaum.

**sufficient**   $P$ is **sufficient** for $Q$ if $P$ implies $Q$. One also says $P$ **suffices** for $Q$. The idea behind the word is that to know that $Q$ is true *it is enough* to know that $P$ is true. Examples are given under **conditional assertion**. *Citation:* (205).

**superscript**   A string of **characters** is a **superscript** to a character if the string is placed immediately after the character and raised above the base line of the text.

***Example 1***   In the expression $x^{23}$, the string 23 is a superscript to $x$.
   Superscripts are heavily **overloaded** in mathematical usage. They are used:
   a) To indicate a multiplicative **power** of a number or a function. *Citations:* (75), (237), (335).

b) To indicate repeated **composites** of a function. Thus $f^2(x)$ means $f(f(x))$. It perhaps could be used to mean $f(x)f(x)$, but I have not found a citation for that usage. It appears to me that $f(x)f(x)$ would customarily be written $(f(x))^2$. *Citation:* (97).

c) As an index. A superscript used as an index may indicate contravariance. *Citation:* (140).

d) To indicate the **domain** of a function space. *Citation:* (345).

e) To indicate the dimension of a space. *Citation:* (432).

f) To indicate the application of an operator. This is common with the notation $T$ for transpose. *Citations:* (211), (386).

g) As a bound on an operator, for example summation, product and integral. *Citations:* (293), (152).

h) A few authors use a superscript to the *left* of the base character, as in $^{23}x$. This may be an index or have some other specially defined meaning. *Citation:* (405).

**Difficulties** A serious confusion in lower level college math courses occurs between $f^{-1}$ as the reciprocal of a function and $f^{-1}$ as the inverse. This is a special case of items (a) and (b) above.

**Remark 1** Sometimes in my classes students give answers that show they think that the Cartesian power $\{1,2,3\}^2$ (which is $\{1,2,3\}\times\{1,2,3\}$) is $\{1,4,9\}$.

**suppose** Discussed under let.

**suppression of parameters** An **identifier** or other mathematical **notation** may omit a **parameter** on which the meaning of the notation depends.

**Example 1** A common form of suppression of parameters is to refer to a **mathematical structure** by its **underlying set**. Thus a **group** with underlying set $G$ and **binary operation** $*$ may be called $G$, so that the notation omits the binary operation. This is also an example of **synecdoche**.

"IT'S A BIRD! IT'S A PLANE! IT'S SUPERSCRIPT!"

239

***Example 2***   A parameter that is suppressed from the notation may or may not be announced explicitly in the text. For example, a text may, by the expression $\log x$, refer to the logarithm with base $e$, and may or may not announce this fact explicitly. Note that this is *not* an example of synecdoche.

See also abuse of notation and orthogonal.

**surjective**   A function $f : A \to B$ is **surjective** if for every element $b$ of $B$ there is an element $a$ of $A$ such that $f(a) = b$. One also says $f$ is **onto** $B$. *Citations:* (65), (255), (392), (403).

***Remark 1***   In view of the differences in the way function is defined mathematically (discussed under function), one should strictly speaking either adopt the stance that every function is equipped with a codomain, or one should always attach a phrase of the form "onto $B$" to any occurrence of the word "surjective".

***Remark 2***   A phrase such as "$f$ is a map onto $S$" does not always mean it is surjective onto $S$. *Citation:* (303).

See also trivial and the sidebar under injective.

**symbol**   A **symbol** is an identifier used in the symbolic language which is a minimal arrangement of characters. "Minimal" means it is not itself constructed of (mathematical) symbols by the rules of construction of symbolic expressions.

***Example 1***   The symbol for the ratio between the circumference and the diameter of a circle is "$\pi$". This is a mathematical symbol consisting of one character.

***Example 2***   The symbol for the sine function is sin. This is a symbol made up of three characters. Although one of the characters is $i$ and $i$ is itself a symbol, its role in the symbol "sin" is purely as a character. (One could think "sin" is the product of $s$, $i$ and $n$, and indeed students do sometimes assume such things, but that would not be the author's intent.)

This is in contrast to the role of $i$ in the **symbolic expression** $3i^2$, a compound expression (not called a symbol in this Handbook) whose meaning is determined **synthetically** by the meanings of the symbols 3, $i$ and 2 and the way they are arranged.

**Remark 1**   Many authors, for example [Fearnley-Sander, 1982] and [Harel and Kaput, 1992], use "symbol" to mean what I call **symbolic expression**. Others use "symbol" to mean **character**.

**Remark 2**   The syntax of symbols and symbolic expressions in the **mathematical register** needs analysis. It appears to me that they are treated like proper nouns. In particular, they don't take the **article**.

**Example 3**   Compare "Sym$_3$ is noncommutative" with "Flicka is a horse".

**Example 4**   Symbols are used in apposition like proper nouns. Two nouns are in **apposition** if one follows the other and they play the same syntactic role. Usually the second is a specification or explanation of the first, as in "My friend Flicka". Compare "The group Sym$_3$" This applies to **variables** as well as **determinate** symbols, as in "the quantity $x^2 + 1$" and "for all integers $n$".

See also **name**.

*References:* This discussion derives in part from [de Bruijn, 1994], page 876.

**symbolic assertion**   See **assertion**.

**symbolic expression**   A **symbolic expression** (or just **expression**) is a collection of mathematical **symbols** arranged

  a) according to the commonly accepted rules for writing mathematics, or

241

b) according to some mathematical definition of a formal language. Every expression is either a **term** or an **symbolic assertion**. In particular, every symbol is a symbolic expression.

The meaning of a symbolic expression is normally determined synthetically from the arrangement and the meanings of the individual symbols. See **semantics** and **compositional**.

**Example 1**   The expressions $x^2$ and $\sin^2 \pi$ mentioned under **symbol** are symbolic expressions. "$\sin^2 \pi$" is an arrangement of three **symbols**, namely sin, 2 and $\pi$. The arrangement itself is meaningful; "$\sin^2 \pi$" is not the same symbolic expression as $2 \sin \pi$ even though they **represent** the same **mathematical object**.

**Remark 1**   As the example indicates, the "arrangement" need not be a string.

*Citations:* (93), (286), (316), (351), (407).

**Subexpressions**   An expression may contain a **subexpression**. The rules for forming expressions and the use of **delimiters** allow one to determine the subexpressions.

**Example 2**   The subexpressions in $x^2$ are $x^2$, $x$ and 2. Two of the subexpressions in $(2x + 5)^3$ are $2x$ and $2x + 5$. The rules of algebra require the latter to be inclosed in parentheses, but not the former.

**Example 3**   Is $\sin \pi$ a subexpression of $\sin^2 \pi$? This depends on the rules for construction of this expression; there is no book to consult because the rules for symbolic expressions in the mathematical register are not written down anywhere, except possible in the bowels of Mathematica®(see Remark 3 under **symbolic language**). One could imagine a rule that constructs the function $\sin^2$ from sin and notation for the squaring function, in which case $\sin \pi$ is not a subexpression of $\sin^2 \pi$. On the other hand, one could imagine a system in which one constructs $(\sin \pi)^2$

and than a Chomsky-style transformation converts it to $\sin^2 \pi$. In that case $\sin \pi$ is in some sense a subexpression of $\sin^2 \pi$.

This example shows that determining subexpressions from the typographical arrangement is not a trivial task. One must understand the rules for forming expressions, implicitly if not explicitly.

**Example 4**   The set

$$\{f \mid f = \sin^n, n \in \mathbb{N}, n > 0\}$$

could also be written

$$\{f \mid f \text{ is a positive \textbf{integral} power of the sine function}\}$$

showing that English phrases can occur embedded in symbolic expressions.

*References:* Symbols and symbolic expressions are discussed in the context of mathematical education in [Schoenfeld, 1985], [Harel and Kaput, 1992], [Tall, 1992c].

**symbolic language**   The **symbolic language** of mathematics is a distinct special-purpose language used independently and in phrases included in **discourse** in the **mathematical register**. It consists of **symbolic expressions** written in the way mathematicians traditionally write them. They may stand as complete sentences or may be incorporated into statements in English. Occasionally statements in English are embedded in symbolic expressions. (See Remark 2 under **identifier**.)

**Example 1**   "$\pi > 0$." This is a complete **assertion** (**formula**) in the symbolic language of mathematics.

**Example 2**   "If $x$ is any **real number**, then $x^2 \geq 0$." This is an **assertion** in the **mathematical register** containing two symbolic expressions. Note that "$x$" is a **term** and "$x^2 \geq 0$" is a **symbolic assertion** incorporated into the larger assertion in English. See **parenthetic assertion**.

***Example 3***  "$\{n \mid n \text{ is even}\}$." This is a **term** in the symbolic language that contains an embedded phrase in the **mathematical register**.

***The symbolic language as a formal structure***  The symbolic language of mathematics has never been given by a mathematical **definition** (in other words it is not a **formal language**). There would be difficulties in doing so.

- The symbolic language has **irregular syntax** and is probably **context-sensitive**.
- The symbolic language of mathematics has many variants depending on the field and individual idiosyncrasies.
- Even if one gives a formal definition one would have problems with mechanical parsing (automatically computing the meaning of an expression) because the language contains **ambiguities**. Examples:
  - Is "$ma$" a **symbol** or is it $m$ times $a$? (See the sidebar under **variate**.)
  - Is $\sin x$ the result of function application or is it a product of two variables named sin and $x$?
  - What does $\sin^{-1} x$ mean?

Mathematica® 3.0 has a standardized version (`StandardForm`) of the **symbolic expression** language of the **mathematical vernacular** that eliminates ambiguities, and it can also output symbolic expressions in a form called `TraditionalForm` that is rather close to actual usage. (See [Wolfram, 1997], pages 187ff.) Presumably the implementation of `TraditionalForm` would have involved a definition of it as a formal language.

De Bruijn [1994] proposes modeling a large part of the **mathematical vernacular** (not just the symbolic language) using a programming language.

**symbolic logic**   Another name for **mathematical logic**.

**symbolitis**   The excessive use of symbols (as opposed to English words and phrases) in mathematical writing – the meaning of "excessive", of course, depends on the speaker! There seems to be more objection to symbols from mathematical logic such as $\forall$ and $\exists$ than to others. *Reference:* This name was given by [Gillman, 1987], page 7.

**symbol manipulation**   **Symbol manipulation** is the transformation of a **symbolic expression** by using algebraic or syntactic rules, typically with the intention of producing a more satisfactory expression. Symbol manipulation may be performed as a step in a **proof** or as part of the process of solving a problem.

***Example 1***   The proof that $a^2 - b^2 = (a + b)(a - b)$ based on the distributive law, the commutative law for multiplication, and the algebraic laws concerning additive inverses:

$$(a + b)(a - b) = a(a - b) + b(a - b) = a^2 - ab + ba - b^2$$
$$= a^2 - ab + ab - b^2 = a^2 - b^2$$

An example of a proof by symbolic manipulation of formulas in mathematical logic is given under **conceptual**. Proof by symbol manipulation is contrasted with **conceptual** proof. See also **aha**.

***Difficulties***   Students often manipulate symbols inappropriately, using rules not valid for the **objects** being manipulated. This is discussed by Harel [1998].

**synecdoche**   **Synecdoche** is naming something by naming a part of it.

***Example 1***   Referring to a car as "wheels".

***Example 2***   Naming a **mathematical structure** by its **underlying set**. This happens very commonly. This is also a case of **suppression of parameters**.

***Example 3***  Naming an equivalence class by a member of the class. Note that this is not an example of suppression of parameters. See **well-defined**. *Citations:* (20), (150), (186),

See also **metaphor**.

*Reference:* [Presmeg, 1997b].

**syntax**  The **syntax** of an **expression** is an analysis of the manner in which the expression has been constructed from its parts.

***Example 1***  The syntax of the expression $5 + 3$ consists partly of the fact that "5" is placed before "+" and "3" after it, but the syntax is more than that; it also includes the fact that "+" is a **binary operation** written in **infix notation**, so that the expression $5 + 3$ is a **term** and not an **assertion**. The expression $3 + 5$ is a *different* expression; the **semantics** usually used for this expression tells us that it has the same **value** as $5 + 3$.

***Example 2***  The syntax of the expression $5 > 3$ tells us that it is an **assertion**; the semantics tells us that it is a true assertion.

***Example 3***  The syntax of the expression $3x + y$ is different from the syntax of $3(x + y)$. In the common tree notation for syntax the two expressions are parsed as in the diagrams in the margin.

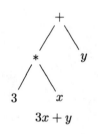

$3x + y$

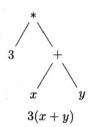

$3(x + y)$

***Using syntax***  The syntax of an expression gives it structure beyond being merely a string of symbols. The structure must be deduced by the reader with clues given by convention (in the case of Example 3, that multiplication takes **precedence** over addition), **parentheses**, and the context. (See also Example 3 under **symbolic expression**.)

Successful students generally have the **functional knowledge** needed to determine the structure without much explicit instruction, and in many cases without much conscious awareness of the process. For example, college students may be able to calculate $3(x+y)$ and $3x+y$ correctly for given **instantiations** of $x$ and $y$, but they may have never consciously noticed that in calculating $3x + y$ you must calculate the product before the sum, and the other way around for $3(x + y)$. (A **reverse Polish** calculator forces you to notice things like that.) See also **compositional**.

The way the order of calculation is determined by the syntactic structure and the observation in section (b) under **substitution** that substitution commutes with evaluation are basic aspects of learning to deal with mathematical expressions that are essentially never made explicit in teaching. (No teacher under whom I studied ever made them explicit.)

Students vary widely on how much they are able to use the syntax to help decode mathematical **expressions**. Even college engineering and science students don't always understand the difference between expressions such as $-2^2$ and $(-2)^2$.

Similarly, the syntax of a complicated English sentence may help some understand it while communicating little or nothing to others. Thus a student may be able to understand a very complicated statement in English that is in context, but will find meaningless a statement with the same logical structure about **abstract objects**.

See also **compositional**, **substitution**.

*Acknowledgments:* Some of this discussion was suggested by [Dubinsky, 1997]. A good reference to the syntax of English is [McCawley, 1988a], [McCawley, 1988b]. Thanks also to Atish Bagchi and Eric Schedler.

**synthetic**   See **compositional**.

**tangent**   The word **tangent** refers both to a straight line in a certain relation to a differentiable curve (more general definitions have been given) and to a certain **trigonometric function**. These two meanings are related: the trigonometric tangent is the length of a certain line segment tangent to the unit circle ($\overrightarrow{DT}$ in the picture).   Sadly, some students get

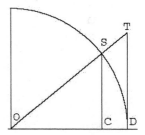

to college without ever knowing this. Similar remarks apply to **secant** ($\overrightarrow{OT}$ in the picture). See **suggestive name**.

**term**   The word **term** is used is several ways in the **mathematical register**.

**(a) A constituent of a sum or sequence**   A term is one constituent of a sum (finite or infinite), in contrast to a **factor**, which is a constituent of a product. The word is also used for a constituent of an infinite sequence.

***Example 1***   The second term in the expression $x^2 + 2x + \frac{1}{x}$ is $2x$.

*Citations:* (8), (258).

| The two usages in (a) and (b) conflict. The expression $x^2 + 1$ is a term in the logical sense, and it is a factor in the expression $(x^2 + 1)(x^2 + 3)$. This can and does lead to confusion when mathematicians and logicians talk. |
| --- |

**(b) A symbolic expression denoting a mathematical object**   In mathematical logic, a term is a symbolic expression that denotes a (possibly **variable**) mathematical object. This is in contrast to a **symbolic assertion**. A term is thus equivalent to a **noun phrase**, whereas an assertion is equivalent to a whole **sentence**.

***Example 2***   Any symbol that denotes a (possibly variable) mathematical object is a term. Thus $\pi$ and 3 are terms.

***Example 3***   The expression $2 + 5$ is a term that denotes 7.

***Example 4***   The expression $x + 2y$ is a term. It denotes a variable number. If specific numbers are substituted for $x$ and $y$ the resulting expression is a term that (in the usual **extensional** semantics) denotes a specific number.

***Example 5***   The expression

$$\int_1^2 x \, dx$$

is a term; it (**extensionally**) denotes the number $3/2$.

*Citations:* (289), (371).

See also formula.

**(c) In terms of** Writing a function **in terms of** $x$ means giving a defining equation containing $x$ as the only variable (but it may contain parameters).

**Example 6** The expression $xy = 1$ implicitly determines $y$ as a function of $x$, explicitly $y$ is given by $y = \frac{1}{x}$ as a function of $x$. *Citation:* (92).

    *Acknowledgments:* Owen Thomas.

**TFAE**   Abbreviation for "the following are equivalent".

# that is

**(a) Indicating equivalence** The phrase "that is" may be used to indicate that what follows is equivalent to what precedes, usually when the equivalence is essentially a rewording.

**Example 1** "We have shown that $xy < 0$, that is, that $x$ and $y$ are nonzero and of opposite sign."

**Example 2** "Then $n = 2k$, that is, $n$ is even."

**(b) Indicating significance** "That is" is sometimes used in the situation that what after is an explanation of the significance of what comes before. The explanation may be a rewording as in (a) above, so that these two usages overlap.

**Example 3** "If the function is a polynomial, that is, easy to calculate, numerical estimates are feasible."

    See also i.e.

    *Citations:* (1), (66), (159), (196), (342),

**the**   See definite article.

**then**    The word **then** in the **mathematical register** generally means that what follows can be deduced from the preceding assumption, which is commonly signaled by if or **when**. See **conditional assertion** and **imply**.

***Example 1***    "If $n$ is divisible by 4, then $n$ is even."

***Remark 1***    Occasionally "then" has a temporal meaning. *Citations:* (28), (79).

**theorem**    To call an **assertion** a **theorem** is to claim that the assertion has been proved. *Citations:* (23), (28), (69), (147).

     Some authors refer only to assertions they regard as important as theorems, and use the word **proposition** for less important ones. See **lemma** and **corollary**. See also **delineated**, **labeled style** and Metaphor (f) under **time**.

**theory of functions**    In older mathematical writing, the phrase **theory of functions** refers by **default** to the theory of analytic functions of one complex **variable**. *Citations:* (417), (419).

**therefore**    **Therefore** means that what follows is a consequence of what precedes, sometimes in a mathematical **argument** and sometimes in a rhetorical argument. It has a connotation of presenting a big, important conclusion that "thus" does not have. Sometimes abbreviated at the blackboard by using the symbol "∴". *Citations:* (34), (420).

**thus**    Like "therefore", **thus** means that what follows is a consequence of what precedes. It may also indicate that what follows is an example of what precedes. *Citations:* (26), (203), (269).

**tilde**    The symbol "˜" is used
- over a letter to create a new variable,
- as a relation meaning asymptotically equal, and
- in web addresses.

The symbol is pronounced "tilde" (till-day or till-duh), or informally "twiddle" or "squiggle". Thus "$\tilde{x}$" is pronounced "x tilde", "x twiddle" or "x squiggle". See also **accent**. *Citations:* (30), (178), (405).

**time**    The concept of **time** is used in several ways in **mathematical discourse**.

### (a) Reference to actual time

***Example 1***    "Fermat stated his famous 'Last Theorem' in about 1630."

One also refers to things happening in time when a program is run.

***Example 2***    "A local variable in a procedure disappears when the procedure is finished." *Citations:* (18), (92), (205), (339).

### (b) Model time by a variable
"The velocity is the derivative of the position with respect to time." *Citations:* (62), (205).

### (c) Metaphor: Variation thought of as taking place in time
A variable may be visualized as varying over time even if there is no stated application involving time.

***Example 3***    "We find a maximum by varying $x$ until $y$ stops going up and starts going down."

***Example 4***    "The function $f(x) = x^2 - 1$ **vanishes** when the function $g(x) = x - 1$ vanishes, but not conversely." *Citations:* (61) (331) (421)

One special case of this is the common picture of moving a geometric figure through space.

***Example 5***    Revolve the curve $y = x^2$ around the $y$-axis. *Citation:* (349).

This metaphor is presumably behind the way the words **increasing** and **decreasing** are used.

### (d) Metaphor: Run through all instances
One thinks of taking the time to check through all the instances of a **structure** denoted by a **variate identifier**.

251

***Example 6***   "When is the integral of a function expressible as a **rational** function?" (The **metaphor** here is that we look through all the functions and note which ones have integrals that are rational functions.)

   *Citations:* (59), (72), (114).

***Example 7***   "When does a symmetric **group** have a nontrivial center?" (Metaphor: Look at all the symmetric groups with an eye out for nontrivial centers.)

***(e) Metaphor: Implication takes time to happen***   A conditional assertion "If $P$, then $Q$" uses "then" in a **metaphor** that probably comes from thinking of $P$ as *causing* $Q$ to be true, and the effect of making $Q$ true takes time to happen. This metaphor is probably the reason "if" is often replaced by "when". (See examples under **when**).

***(f) Metaphor: Progressing through mathematical discourse***   In a text, the past tense may refer to part of the text above the reference; the future tense may refer to a part below it. (Note that I had to use another metaphor – that text is a vertical column – to say this.) Other references to time are used in this way, too.

***Example 8***   "We proved the **forward** part of the theorem in Chapter 2 and will prove the converse part in Chapter 5." (In real time, of course, the converse part has *already* been proved.)

***Example 9***   "From now on we shall denote the binary operation by **juxtaposition**".

***Example 10***   "For the moment, suppose $x > 0$." *Citations:* (5), (12), (45), (94), (96), (333), (355).

***(g) Metaphor: Progress through a calculation or a proof***

***Example 11***   "We iterate this construction to obtain all finite binary trees."

Progressing through a calculation is not included under Metaphor (f) because the intent of the author is often that one *imagines* going through the calculation; the calculation is not actually exhibited in the discourse.

*Example 12* "Every time we perform the elimination step $n$ is reduced in value, so this calculation must eventually stop."

*Example 13* "The starting point for our proof is the observation that the sum of two even integers is even." *Citations:* (36), (34), (57), (130).

*Remark 1* This may be the metaphor behind the fact that a **quick proof** means a short proof.

See also **always, never, location.**

**translation problem** The **translation problem** is the name used in this Handbook for the process of discovering the logical structure of a mathematical **assertion** $P$ that has been stated in the **mathematical register**. This is essentially equivalent to the problem of finding an **assertion** in **mathematical logic** that represents $P$. Learning how to do this is one of the difficult skills students of mathematics have to acquire, even very early with simple word problems.

Many of the entries in this Handbook illustrate the complications this involves; see for example **and, conditional assertion** and **universal quantifier.**

*References:* This is discussed in the context of **mathematical education** in [Selden and Selden, 1999], where discovering the logical structure of an assertion is called **unpacking**. The text [Kamp and Reyle, 1993] is essentially a study of the analogous problem of discovering the logical structure of statements in ordinary English rather than in the **mathematical register.**

**trial and error** The process of finding an **object** that satisfies certain **constraints** by guessing various possibilities and testing them against the constraint until you find one that works is called the process of **trial and error**. See Example 1 under **algorithm addiction**. Trial and error is valid when it works but it typically takes exponential time in the size of the problem to carry out. *Citations:* (348), (232).

## trigonometric functions

*(a) Degrees and radians* It is not always explicitly noted to students that if you write $\sin x$ meaning the sine of $x$ degrees, you are not using the same function as when you write $\sin x$, meaning the sine of $x$ radians. They have different derivatives, for example. The same remark may be made of the other trigonometric functions. This point is correctly made in [Edwards and Penney, 1998], page 167.

It appears to me that in postcalculus pure mathematics "$\sin x$" nearly always refers to the sine of $x$ radians (not degrees), often without explicitly noting the fact. This is certainly not true for texts written by non-mathematicians, but the situation is made easier by the customary use of the degree symbol when degrees are intended.

*(b) Variations in usage* In the USA one calculates the sine function on the unit circle by starting at $(1, 0)$ and going counterclockwise (the sine is the projection on the $y$-axis), but in other countries one may start at $(0, 1)$ and go clockwise (the sine is the projection on the $x$-axis). I learned of this also from students, but have been unable to find citations.

Students educated in Europe may not have heard of the secant function; they would simply write $\cos^{-1}$.

*(c) Evaluation* One normally writes evaluation of trigonometric functions by juxtaposition (with a small space for clarity), thus $\sin x$ instead of $\sin(x)$. Students may sometimes regard this as multiplication.

See also logarithm and tangent.

*Acknowledgments:* Michael Barr.

## trivial

*(a) About propositions* A fact is said to be **trivial** to prove if the fact follows by **rewriting using definitions**, or perhaps if the common **mental representation** of the **mathematical objects** involved in the fact makes the

254

truth of the fact immediately perceivable. (This needs further analysis. I would tend to use "obvious" for the second meaning.)

***Example 1*** A textbook may define the **image** of a **function** $F : A \to B$ to be the set of all elements of $B$ of the form $F(a)$ for some $a \in A$. It then goes on to say that $F$ is **surjective** if for every element $b$ of $B$ there is an element $a \in A$ with the property that $F(a) = b$. It may then state a theorem, or give an exercise, that says that a function $F : A \to B$ is surjective if and only if the image of $F$ is $B$. The **proof** follows immediately by **rewriting using definitions**.

I have known instructors to refer to such an assertion as "trivial" and to question the worth of including it in the text. In contrast, I have known *many* students in discrete math and **abstract algebra** classes who were totally baffled when asked to prove such an assertion. *This disparity between the students' and the instructors' perception of what is "trivial" must be taken seriously.*

See **ratchet effect**. *Reference:* [Solow, 1995] is one text with a discussion of image and surjective as described in Example 1.

***(b) About mathematical objects*** A function may be called **trivial** if it is the identity function or a constant function, and possibly in other circumstances. *Citation:* (37).

A solution to an equation is said to be **trivial** if it is the identity element for some operation involved in the equation. There may be other situations in which a solution is called "trivial" as well. *Citations:* (259), (346).

A mathematical structure is said to be **trivial** if its underlying set is empty or a singleton set. In particular, a subset of a set is **nontrivial** if it is nonempty. See **proper**. *Citation:* (64).

I suspect that teachers (and hotshot math majors) telling students that an **assertion** is "obvious" or "trivial" is an important cause (but not the only one) of the feeling much of the American population has that they are "**bad at math**". In many cases a person who feels that way may have simply not learned to **rewrite using definitions**, and so finds some proofs impossibly difficult that their instructor calls "trivial".

255

***Remark 1*** "Trivial" and **degenerate** overlap in meaning but are not interchangeable. A search for more citations might be desirable, but it is not clear to me that there is a consistent meaning to either word.

**true**   An **assertion** $P$ in **mathematical discourse** is **true** if it can be proved using accepted axioms.

***Remark 1***   The two words "true" and **valid** are distinguished in **mathematical logic**, in which $P$ is true roughly speaking if it can proved by a sequence of deductions from the axioms currently being assumed, and it is valid if it is a correct statement about every **model** of those axioms. The completeness theorem in **first order logic** asserts that these two concepts are the same for statements and proofs in first order logic.

**turf**   If you are defensive about negative comments about your field, or annoyed when another department tries to teach a course you believe belongs in mathematics, you are protecting your **turf**. The use of this word of course is not restricted to mathematicians (nor is the phenomenon it describes).

***Example 1***   I have occasionally witnessed irritation by people familiar with one field at the use of a term in that field by people in a different field with a different meaning. This happened on the `mathedu` mailing list when some subscribers started talking about **constructivism** with the meaning it has in mathematical education rather than the (unrelated) meaning it has in mathematical logic.

**twiddle**   See tilde.

**two**   In **mathematical discourse**, two **mathematical objects** can be one object. This is because two identifiers can have the same value unless some word such as **distinct** is used to ensure that they are different.

**Example 1** "The sum of any two even integers is even". In this statement, the two integers are allowed to be the same.

*Acknowledgments:* Susanna Epp.

**type**   The **type** of a **symbol** is the kind of value it is allowed to have in the current **context**.

**Example 1**  In the **assertion**

"If $f$ is differentiable and $f'(x) = 0$ then $x$ is a critical point of $f$."

we may deduce that $f$ is of type **function** and $x$ is (probably) of type **real**, even if the author does not say this.

This sort of type deduction requires both mathematical knowledge and knowledge of **conventions**; in the present example one convention is that complex numbers are more commonly written $z$ instead of $x$. Mathematical knowledge (as well as convention) tells us that $x$ cannot be of type integer.

> Mathematicians do not use the word "type" much in the sense used here. When they do use it it typically refers to a classification of structures in a particular field, as in for example differential equations of hyperbolic type.

***Types and sets***  One could dispense with the concept of type and refer to the **set** of possible values of the symbol. It appears to me however that "type" is psychologically different from "set". Normally one expects the type to be a natural and homogeneous kind such as "function" or "real number", not an **arbitrary** kind such as "real number bigger than 3 or integer divisible by 4". One has no such psychological constraint on sets themselves. This needs further investigation.

***Type labeling***  **Type labeling** means giving the **type** of a **symbol** along with the symbol. It is a form of **redundancy**.

**Example 2**  If it has been established on some early page of a text that $S_3$ denotes the symmetric group on 3 letters. A later reference to it as "the group $S_3$" or "the symmetric group $S_3$" is an example of type labeling.

**Remark 1**    Russian mathematical authors seem to do this a lot, although that may be because one cannot attach grammatical endings to symbols.

    *References:* Jeffrey Ullman, in a guest appearance in [Knuth, Larrabee and Roberts, 1989], flatly recommends *always* giving the type of a symbol. Using explicit typing in teaching is advocated in [Wells, 1995]. See also [Bagchi and Wells, 1998a] and [Lamport, 1992].

---

The discussion here is about using the concept of type in communicating mathematics, not its use in logic or foundations. It is not inconsistent to believe both of these statements:

- In mathematical teaching and writing, it is helpful to mention the type of a variable.
- Formal mathematical logic is best done using untyped variables.

---

**Difficulties**    Students commonly make type mistakes.

- They may talk about $2\pi$ being divisible by 2.
- They may write $A/B$, where $A$ and $B$ are matrices.

It is helpful to refer to the concept explicitly as a way of raising consciousness. This is discussed in [Wells, 1995]. *Citations:* (194), (248).

**under**    Used to name the **function** by which one has computed the value, or the function being used as an operation.

**Example 1**    "If the value of $x$ under $F$ is greater than the value of $x$ under $G$ for every $x$, one says that $F > G$." *Citation:* (139).

**Example 2**    "The set $\mathbb{Z}$ of integers is a **group** under addition." *Citations:* (68), (284).

**Example 3**    "If $x$ is related to $y$ under $E$, we write $xEy$." *Citation:* (128).

    See **law of gravity for functions**.

**underlying set**    See **mathematical structure**.

**understand**    Good students frequently complain that they can do the calculations but they don't understand the concepts. (Poor students can't do the calculations!) What they are missing are useful **metaphors** with which they can think about the **mathematical objects** involved in their calculations. This is discussed further under **metaphor**.

*References:* Much new insight has been gained in recent years by cognitive scientists and researchers in **mathematical education** concerning what it means to understand something. A good way to get into the literature on this subject is to read [Lakoff and Núñez, 2000] and the works of Sfard, particularly [Sfard, 1994] and [Sfard, 1997]. She gives many references.

**unique** To say that an **object** satisfying certain **conditions** is **unique** means that there is only one object satisfying those conditions.

*Citations:* (66), (78), (138), (391). As some of the **citations** show, the unique object may be a variable object determined uniquely by explicit parameters.

This meaning can have philosophical complications.

*(a) Is a natural number unique?* The statement in Example 3 under **mathematical object** that 3 is a **specific mathematical object** would not be accepted by everyone. As Michael Barr pointed out in a response to a previous version of this entry, there are various possible definitions of the natural numbers and each one has its own element called 3. (See **literalist**.) Nevertheless, mathematicians normally speak and think of the number 3 as one specific mathematical object, and it is customary usage that this Handbook is concerned with.

*(b) Is the set of natural numbers unique?* The phrase "the set of **natural numbers**" causes a similar problem. This phrase could **denote** any of the standard models of the Peano axioms, all of which are **isomorphic**. Some mathematicians would say that the natural numbers are "unique **up to** isomorphism". Others would say, "I choose one of the models and that will be the natural numbers." Still others would simply assert that there is a *unique* set of natural numbers and all the talk above is about

foundations, not about what mathematicians actually deal with. Most mathematicians in ordinary discourse speak of the natural numbers as if they were unique, whatever they believe.

*(c) Unique up to isomorphism* "The symmetric group on $n$ letters" is unique up to isomorphism, but in contrast to the Peano natural numbers, it is not unique up to a *unique* isomorphism. Algebraists may nevertheless talk about it as if it were unique, but when pressed by literal-minded listeners they will admit that it is determined only up to isomorphism.

    The word "unique" is misused by students; see in your own words. See also up to.

**universal generalization**   If you have proved $P(c)$ for a variable object $c$ of some type, and during the proof have made no restrictions on $c$, then you are entitled to conclude that $P(x)$ is true for all $x$ of the appropriate type. Thus to prove that the square of an even integer is even, one would start with, "Let $n$ be even ... " and then show without making further assumptions about $n$ that $n^2$ is even. This process is formalized in mathematical logic as the rule of deduction called **universal generalization**.

**universal instantiation**   If it is known that $P(x)$ is true of all $x$ of the appropriate type, and $c$ is the identifier of a specific mathematical object of that type, then you are entitled to conclude that $P(c)$ is true. In mathematical logic, the formal version of this is known as **universal instantiation**.

**universal quantifier**   An expression in mathematical logic of the form $\forall x\, P(x)$, where $P$ is a predicate, means that $P(x)$ is true for every $x$ of the appropriate type. The symbol $\forall$ is pronounced "for all" and is called the **universal quantifier**.

*Expressing universal quantification in the mathematical register*
When a universally quantified sentence in the mathematical register is
translated into a sentence of the form $\forall x\, P(x)$ in mathematical logic, the
assertion $P(x)$ is nearly always in the form of a conditional assertion.
Thus in particular all the sentences listed as examples under conditional
assertion provide ways of expressing universal quantification in English.
However, there are other ways of doing that that are not conditional
assertions in English. To provide examples, let $C(f)$ mean that $f$ is
continuous and and $D(f)$ mean that $f$ is differentiable. The assertion
$\forall n\,(D(n) \Rightarrow C(n))$ can be said in the following ways:

a) **Every** differentiable function is continuous. *Citations:* (28), (307).

b) **Any** differentiable function is continuous. *Citations:* (119), (314).

c) **All** differentiable functions are continuous. *Citations:* (7), (91),
   (108), (322).

d) Differentiable functions are continuous. *Citation:* (228).

e) A differentiable function is continuous. *Citations:* (23), (273), (378).

f) Each differentiable function is continuous. *Citations:* (59), (66),
   (104), (343), (369).

g) **The** differentiable functions are continuous. (This sounds like an
   obsolescent usage to me.) *Citation:* (19).

One can make an explicit conditional assertion using the same words:

h) For every function $f$, if $f$ is differentiable then it is continuous.
   *Citations:* (46), (410).

i) For any function $f$, if $f$ is differentiable then it is continuous.
   *Citation:* (71), (54).

j) For all functions $f$, if $f$ is differentiable then it is continuous. *Citations:* (91), (335).

In any of these sentences, the "for all" phrase may come after the
main clause. The conditional assertion can be varied in the ways described
under that listing. See also **each**.

261

If the variable is **typed**, either the **definite** or the **indefinite article** may be used:

k) "If the function $f$ is differentiable, then it is continuous."

l) "If a function $f$ is ... ".

   *Citation:* (342).

**Remark 1**   Sentences such as (d), (e) and (g) are often not recognized by students as having universal quantification force. Sentence (e) is discussed further under **indefinite article**, and sentence (f) is discussed further under **each**.

See also **always**, **distributive plural** and **negation**.

**Universal quantification in the symbolic language**   The quantifier is sometimes expressed by a constraint written to the right of a displayed **symbolic assertions**.

**Example 1**   The assertion, "The square of any real number is nonnegative" can be written this way:

$$x^2 \geq 0 \qquad\qquad \text{(all real } x)$$

or less explicitly

$$x^2 \geq 0 \qquad\qquad (x)$$

One might write "The square of any nonzero real number is positive" this way:

$$x^2 > 0 \qquad\qquad x \neq 0$$

*Citation:* (79).

**Open sentences**   Sometimes, the quantifier is not reflected by any **symbol** or English word. The sentence is then an **open sentence** and is interpreted as universally quantified. The clue that this is the case is that the **variables** involved have not in the present **context** been given specific values. Thus in [Grassman and Tremblay, 1996], page 105:

262

"A function $f$ of arity 2 is *commutative* if $f(x,y) = f(y,x)$."
This means that $f(x,y) = f(y,x)$ for all $x$ and all $y$.

**Remark 2**  Sometimes an author does not make it clear which variable is being quantified.

" In fact, every $Q_i(s) \cong 1 \pmod{m}$, since.... "

The context shows that this means

$$\forall i \left( Q_i(s) \cong 1 \pmod{m} \right)$$

(This is from [Neidinger and Annen III, 1996], page 646.)

**Difficulties**  Students sometimes attempt to prove a **universally quantified assertion** by giving an example. They sometimes specifically complain that the instructor uses examples, so why can't they? There are several possibilities for why this happens:

- The students have seen the instructor use examples and don't have a strong sensitivity to when one is carrying out a proof and when one is engaged in an illuminatory discussion.

- The student has seen **counterexamples** used to disprove universal statements, and expects to be able to prove such statements by a kind of **false symmetry**.

- The student is thinking of the example as generic and is carrying out a kind of **universal generalization**.

- The problem may have expressed the universal quantifier as in Example 1 under **indefinite article**.

*Acknowledgments:* Atish Bagchi, Michael Barr.

*References:* [Epp, 1999]. The texts [Exner, 2000], chapter 3, [Wood and Perrett, 1997], page 12 are written for students. For studies of quantification in English, see [Chierchia and McConnell-Ginet, 1990] and [Keenan and Westerståhl, 1997].

See also **always**, **counterexample**, **never**, **existential quantifier** and **order of quantifiers**.

263

**unknown**   One or more variables may occur in a constraint, and the intent of the discourse may be to determine the values of the variables that satisfy the constraint. In that case the variables may be referred to as **unknowns**.

**Remark 1**   The variable is commonly numerical and the constraint is commonly an equation, but the word occurs in other contexts as well, for example finding the unknown function (or parametrized family of functions) satisfying a differential equation.

Determining the "values of the variables that satisfy the constraint" may mean finding the shape determined by the constraints (for example, the unit circle determined by $x^2 + y^2 = 1$.)

*Citations:* (17), (43), (215), (408).

**unnecessarily weak assertion**   Students are often uncomfortable when faced with an assertion such as

" Either $x > 0$ or $x < 2$ "

because one could obviously make a stronger statement. The statement is nevertheless true.

**Example 1**   Students have problems both with "$2 \leq 2$" and with "$2 \leq 3$". This may be compounded by problems with inclusive and exclusive or.

**Remark 1**   It appears to me that unnecessarily weak statements occur primarily in these contexts:
  a) When the statement is what follows formally from the preceding argument.
  b) When the statement is made in that form because it allows one to deduce a desired result.

I believe students are uncomfortable primarily in the case of (b), and that their discomfort is an instance of **walking blindfolded**.

*Acknowledgments:* Michael Barr.

**unwind**  A typical definition in mathematics may make use of a number of previously defined concepts. To **unwind** or **unpack** such a definition is to replace the defined terms with explicit, spelled-out requirements. This may change a **conceptual** definition into an **elementary** definition. An example is given under **elementary**. See **rewrite using definitions**. *Citations:* (38), (202), (227).

**up to**  Let $E$ be an equivalence relation. To say that a definition or description of a type of **mathematical object** determines the object **up to** $E$ (or modulo $E$) means that any two objects satisfying the description are equivalent with respect to $E$.

**Example 1**  An indefinite integral $\int f(x)\,dx$ is determined up to a constant. In this case the equivalence relation is that of differing by a constant.

The objects are often described in terms of parameters, in which case any two objects satisfying the description are equivalent once the parameters are **instantiated**.

**Example 2**  The statement "$G$ is a finite **group** of order $n$ containing an element of order $n$" forces $G$ to be the cyclic group of order $n$, so that the statement defines $G$ up to **isomorphism** once $n$ is instantiated.

See **copy**. *Citations:* (146), (274).

**uppercase**  See **case**.

**vacuous implication**  A conditional assertion "If $A$ then $B$" is **true** if $A$ happens to be false. This is not usually the interesting case, so this phenomenon is called **vacuous implication**.

**Difficulties**  Students have a tendency to forget about vacuous implication even if reminded of it.

**Example 1**  A relation $\alpha$ is **antisymmetric** if
$$(a\ \alpha\ b)\ \text{and}\ b\ \alpha\ a) \Rightarrow a = b$$
The relation "$<$" on the set of all reals is antisymmetric. When this is pointed out, a student may ask, "How can less-than be antisymmetric? It's impossible to have $r < s$ and $s < r$!"

**valid**  In common mathematical usage, an assertion $P(x)$ is **valid** if it is correct, that is, if for all $x$ of the correct type, $P(x)$ is **true**. Thus valid and true mean essentially the same thing in common usage.

A **proof** is valid if all its steps are correct.

**value**  The **object** that is the result of evaluating a **function** at an element $x$ of its **domain** is called the **value**, **output** or **result** of the function at $x$. *Citations:* (133), (212), (289).

The word "value" is also used to refer to the **mathematical object** denoted by a literal expression. Most commonly the word is used when the value is a **number**. *Citations:* (302), (323).

**(a) *Symbolic notation for value***  If the function is denoted by $f$, then the value at $x$ is denoted by $f(x)$ or $fx$ (**prefix notation**). Whichever notation an author routinely uses, the value at $x + 1$ would of course be denoted by $f(x + 1)$. For an author who always writes $f(x)$ (except for function symbols that normally don't use parentheses – see **irregular syntax**), the **parentheses** serve to **delimit** the input to the function. For those who normally write $fx$, the parentheses in $f(x+1)$ are used as **bare delimiters**. Both notations have a long history. *Citations:* (45), (164), (407).

The value may be denoted in other ways:
a) $xf$ (**reverse Polish notation**). (The notation $fx$ is a special case of **Polish notation**).

b) $(x)f$ (**postfix notation**). These are discussed in their own entries with examples. See also **rightists**.

c) $f_x$ (mostly for functions defined on **integers**— see **subscript**).

d) $f[x]$. This notation is used by Mathematica®; parentheses are reserved for **grouping**.

See **parentheses** for more about their usage with function values. *Citations:* (313), (317).

More elaborate possibilities exist for functions with more than one input. See **infix notation**, **prefix notation**, **postfix notation**, **outfix notation**.

*(b) Properties of function values* Adjectives applied to a function may refer to its outputs. *Examples:* The phrase **real function** means that the outputs of the function are real (but some authors would prefer "real-valued function"), and similarly for **complex function**. In neither case does the phrase imply that the *domain* has to consist of real or complex numbers. "$F$ is a **positive function**" means that $F(x) > 0$ for every $x$ in its domain. The phrase "positive-valued function" seems to be rare.

In contrast, a **rational function** is a function whose **defining expression** is the quotient of two polynomials. A rational function $F : \mathbb{R} \to \mathbb{R}$ will have rational outputs at rational inputs, but will not have rational output in general. *Citations:* (42); (164), (335), (425).

*(c) Terminological conventions* For many functions, one says that the result of applying the function $f$ to the input $c$ is $f(c)$. For example, the phrase in the non-symbolic part of the **mathematical register** for $\sin x$ is "sine of $x$". However, many operations have a name that is *not* used for the result, which requires another name. In some cases the symbol used has a

| function | symbol name | result |
|---|---|---|
| addition | plus | sum |
| subtraction | minus | difference |
| multiplication | times | product |
| division | divided by | quotient |
| squaring | squared | square |
| composition | composed with | **composite** |
| differentiation | | derivative |
| integration | | integral |

third name. For example the result of *adding* 3 and 5 is $3 + 5$, read 3 *plus* 5, and that is 8, which is the *sum* of 3 and 5. Some common operations for which this holds are listed in the sidebar on page 267.

"Symbol name" refers to the way the symbol is read in speech; thus $a - b$ is read "$a$ minus $b$". Note that both differentiation and integration involve several different symbolic notations.

**Remark 1** These usages are not completely parallel. For example, one can say "$g$ composed with $f$" or "the composite of $g$ and $f$", and similarly "the derivative of $f$", but one cannot say "the plus of 3 and 5". On the other hand, "plus", "minus" and "times" may be used with "sign" to name the symbol directly, but the symbol "$\div$" is called the "division sign", not the "divided by" sign, and there is apparently no common name for the sign for composition.

**Remark 2** Many writers blur the distinction between composition and composite and refer to $g \circ f$ as the "composition" of $g$ and $f$. I have heard students blur the distinction for some of the other operations, as well, for example saying, "8 is the addition of 3 and 5". *Citations:* (43) (78), (80), (87), (114), (198), (286), (229), (302), (326), (327), (337), (350).

**vanish** A function $f$ **vanishes** at an input $a$ if $f(a) = 0$.

**Example 1** "Consider the collection of all continuous functions that vanish at 0." *Citations:* (43), (61), (312).

**variable** The noun **variable** in mathematical discourse generally refers to a **variate symbol**. The word is used primarily in certain **conventional** settings, primarily as a named element of the **domain** of a **function**, as a reference to an unnamed function, and in certain technical terms, most notably "random variable". Most other occurrences of variate identifiers would not be called variables, except in **mathematical logic**, where the word is given a technical definition that in effect refers to any **variate** identifier. This Handbook uses the terminology of mathematical logic.

**Example 1** In the **discourse**

" Let $f$ be a **function** for which $f(x) > 0$ when $x > 2$. "

the $x$ and the $f$ are both variate symbols. In common mathematical parlance only $x$ would be called a variable. In the terminology of **mathematical logic**, both $x$ and $f$ are variables.

**Example 2** Consider the discourse:

"Let $G$ be a **group** with identity element $e$ and an element $a$ for which $a^2 = e$. Then $a = a^{-1}$."

The author or speaker may go on to give a **proof** of the claim that $a = a^{-1}$, talking about $G$, $e$, and $a$ with the same syntax used to refer to physical objects and to **specific mathematical objects** such as 3 or the sine function.

Because of the way the proof is written, the writer will appear to have in mind not any specific group, and not all possible groups, but a *nonspecific* or *variable* group. The symbol $G$ is a variable; it *varies over* groups.

**Example 3** In the phrase "A function of $n$ variables" the word refers to the inputs to the function. If the function is given by a formula the variables would normally appear explicitly in the formula.

**Example 4** The **statement**, "Let $x$ be a variable dependent on $t$" has the same effect as saying "Let $f(t)$ be a function", but now the **value** is called $x$ or $x(t)$ instead of $f(t)$. *Citations:* (4), (106), (205), (235), (276), (351), (370).

The *adjective* "variable" is used to say that the object it modifies is a **variable mathematical object**. *Citations:* (89), (396).

There are several ways to *think about* variables and several ways to *formalize* them.

## (a) How to think about variables

*(i) Variable objects*   One way of understanding the symbol $G$ in Example 2 is that it refers to a **variable mathematical object**.

From this point of view, what $G$ refers to is a genuine **mathematical object** just like 3 or $\mathbb{R}$, the difference being that it is an object that is *variable* as opposed to uniquely determined. This is sound: there are various methods using **mathematical logic** or category theory that give a formal mathematical definition of "variable object".

*(ii) Incomplete specification*   There is another point of view about $G$ that I suspect many mathematicians would come up with if asked about this topic: $G$ refers to *any group that meets the* **constraint** (in this case having a nontrivial element whose square is the identity element). It is thus *incompletely specified*. Statements made about it become true if one deduces the statements from the axioms for groups and the given constraints, and not from any other specific properties a group might have.

*(iii) Variable as role*   A metaphor that in some manner incorporates the "incomplete specification" point of view is that a variable is a **role**; in Example 2, $G$ is a *role that can be played by any group* satisfying the constraint. Then the **proof** is like a play or a movie; when it stars a particular group in the role of $G$ it becomes a proof of the theorem about that group. *References:* The idea of role comes from [Lakoff and Núñez, 2000].

## (b) Formalisms for variables

*(i) Logicians' formalism*   In classical logic an **interpretation** of **discourse** such as that in Example 2 assigns a specific **group** to $G$, its identity element to $e$, an element of that group to $a$, and so on. An **assertion** containing identifiers of **variable mathematical objects** is said to be true if it is true in all interpretations. I will call this the **logician's**

**semantics** of variables. For the purposes of giving a **mathematical definition** of the word "variable", it would be reasonable to identify the variable object with the **symbol** in the **formal language** (such as $G$ in the example above) corresponding to it.

Another possibility would be to identify the variable object with the set of all possible interpretations, although to do that correctly would require dealing with the fact that that "set" might actually be a proper class.

*(ii) Categorists' semantics of variable objects* Categorists have another approach to the concept of variable mathematical object. One defines a **theory**, which is a specific category (the theory for groups, for example). The theory contains a specific object $g$. Every group is the value at $g$ of a certain type of functor based on that theory. It is natural to interpret the object $g$ of the theory (or, perhaps better, the entire theory) as the object denoted by the identifier $G$ in Example 2 above.

*References:* The categorical approach is worked out in [Fourman, 1977], [Makkai and Reyes, 1977], [Fourman and Vickers, 1986], [Lambek and Scott, 1986].

*(iii) Specific approaches* The approaches suggested so far are *general* ways of understanding variable objects. Certain *specific* constructions for particular types of variable objects have been known for years, for example the familiar construction of the variable $x$ in the polynomial ring of a field as an infinite sequence that is all 0's except for a 1 in the second place.

Other aspects of variables are discussed under **bound variable, free variable, determinate, substitution, variate, Platonism** and the discussion after Example 2 under **variable mathematical object**. Related concepts are **constant, parameter** and **unknown**.

**variable clash**   A substitution of an expression containing a **free variable** into an expression that contains and **binds** the same literal variable.

***Example 1***   A student must solve an integral $\int_0^9 r^3 A \, dr$, where she knows that $A$ is the area of a certain circle. She therefore rewrites it as $\int_0^9 r^3 \pi r^2 \, dr$; this will give the wrong answer. *Citation:* (291).

**variate**   A free identifier, either in the **symbolic language** or in English, is **variate** if it is intended to refer to a **variable mathematical object**. A variate identifier, at least in intent, has more than one **interpretation** in the universe of **discourse**. These two points of view — the identifier names a variable mathematical object and the identifier has more than one interpretation — are discussed at length under **variable**.

***Example 1***   In the **assertion**, "If the quantity $a$ is positive, then $a^x$ is positive for all real $x$", $x$ and $a$ are both variate. In contrast, in the phrase "the exponential function $a^x$", $a$ is variate but $x$ is not an identifier, it is a **dummy variable**. In this case, in common usage, $x$ is a **variable** and $a$ is a **parameter**.

***Example 2***   In the passage

    "Let $G$ be a **group** with identity element $e$."

"$G$" and "$e$" are variate.

***Example 3***   "Let $G$ be a group and $g \in G$. Suppose the group $G$ is commutative ... ." This illustrates the fact that variable **mathematical structures** are commonly referred to using **definite noun phrases**.

### Some fine points

a) Being determinate or variate is a matter of the current **interpretation**; it is not an inherent property of the **symbol**, even though some symbols such as $\pi$ are **conventionally** determinate and others such as $x$ are conventionally variate. For example, $\pi$ is sometimes used as the name of a projection function.

b) The distinction between determinate and variate is not the same as the grammatical distinction between **definite description** and **indefinite description**. See Example 1 under **definite description**.

c) The distinction between determinate and variate is not the same as the grammatical distinction between common and proper nouns. Indeed, all **symbolic expressions** seem to use **syntax** very similar to that of proper nouns. See Remark 2 under **symbol**.

d) Variate and determinate identifiers are **free** by definition. Asking whether a **bound variable** is variate or determinate does not in any obvious way make sense. See the sidebar under **bound identifier**.

e) Given the passage "Suppose $x$ is a real **variable** and $3x + 1 = 7$", one deduces that $x = 2$. Its use in that sentence is nevertheless variate. The *intent* is that it be a variable. The **conditions** imposed force it to denote just one number. (It is easy to think of examples where, unlike this one, it is very difficult to determine whether the conditions force a unique value.) It is the intent that matters.

***Terminology*** The names "determinate" and "variate" are my own coinages. I felt it important not to use the phrase "variable identifier" because it is ambiguous.

*Acknowledgments:* Owen Thomas.

**vector**    The word **vector** has (at least) three different useful **mental representations**:

- An $n$-tuple.
- A quantity with length and direction.
- An **element** of a vector space.

Of course, the third representation includes the other two, but with some subtleties. For example, to think of an element of an abstract $n$-dimensional vector space as a $n$-tuple requires choosing a basis. There is in general no canonical choice of basis.

---

[ISO, 1982], quoted in [Beccari, 1997], recommends a practice which in my terminology would be: Use upright typographic **characters** for determinate **symbols** and slanted typographical characters for variate symbols. In fact common practice these days seems to be: Use upright characters for multiletter symbols (such as "sin") and slanted characters for symbols consisting of one character. Michael Barr has commented that this enables us to distinguish multiletter symbols from products.

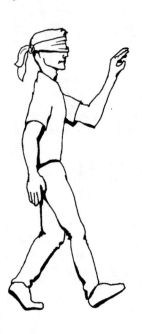

***Variations***  In computer engineering, the word vector is often used to refer to an $n$-tuple of any sort of thing, not necessarily elements of a field, so that the $n$-tuple may indeed not be a member of a vector space.

I have heard this usage in conversation but have not found an unequivocal citation for it.

*Citations:* (76), (172), (321).

Students' understanding of vectors are discussed in the very informative article [Watson, Spirou and Tall, 2003].

**vinculum**  See bar.

**walking blindfolded**  Sometimes a lecturer lists steps in an argument that will indeed culminate in a valid **proof**, but the reason for the steps is not apparent to the student. The student may feel like someone who is walking straight ahead with a blindfold on: how do you know you won't bump into a wall or fall off a cliff? That is **walking blindfolded** (my name). This is closely related to the attitude described in section (a) under **attitudes**.

It is my observation that many students find it difficult or impossible to follow a proof when they cannot see where it is going, whereas others are not bothered at all by this.

See also **look ahead**.

**well-defined**  Suppose you try to define a **function** $F$ on a **partition** $\Pi$ of a set $A$ by specifying its value on an **equivalence** class $C$ of $\Pi$ in terms of an element $x \in C$ (a case of **synecdoche**). For this to work, one must have $F(x) = F(x')$ whenever $x$ is **equivalent** to $x'$. In that case the function $F$ is said to be **well-defined**. (See **radial concept**).

***Example 1***  Let $\mathbb{Z}_2$ be the group of congruence class of integers **mod** 2, with the class of $n$ denoted by $[n]$. Define $F : \mathbb{Z}_2 \to \mathbb{Z}_2$ by $F[n] = [n^2]$. Then $F$ is well-defined (in fact, it is the **identity** function), because an integer is even if and only if its square is even. If you define $G[n]$ to be the number of primes dividing $n$, then $G$ would not be well-defined, since $G[2] = [1]$, $G[6] = [2]$, and $[2] = [6]$ but $[1] \neq [2]$.

Definition on equivalence classes is perhaps the most common use of "well-defined", but there are other situations in which it is used as well.

**Example 2**   Let $\mathbb{P}$ denote the set of all nonempty subsets of the set of nonnegative integers. Define $F : \mathbb{P} \to \mathbb{Z}$ by: $F(A)$ is the smallest element of $A$. Since the nonnegative integers are well-ordered, $F$ is well-defined. This is a case where there might have been doubt that the value exists, rather than worry about whether it is ambiguous.

**Example 3**   Conway and Hofstadter defined a sequence on the positive integers by $a(1) = a(2) = 1$ and

$$a(n) = a(a(n-1)) + a(n - a(n-1))$$

for $n \geq 3$. This is well-defined because one can show by induction that $a(n) < n$ for $n \geq 3$ (otherwise the term $a(n-a(n-1))$ could cause trouble). This example comes from [Mallows, 1991].

See radial concept and literalist.

**Variations**   Many authors omit the hyphen in "well-defined".

   *Citations:* (20), (165).

> There is a subtlety in Example 3. The observation that $a(n) < n$ for $n \geq 3$ does indeed show that the sequence is well-defined, but a sequence can sometimes be well-defined even if the function calls in the definition of the value at $n$ refer to integers larger than $n$. An example is the function
> $$F(n) = \begin{cases} F\Big(F(n+11)\Big) & (n \leq 100) \\ n - 10 & (n > 100) \end{cases}$$

**when**   Often used to mean "if".

**Example 1**   "When a function has a derivative, it is necessarily continuous."

**Remark 1**   Modern dictionaries [Neufeldt, 1988] record this meaning of "when", but the original Oxford English Dictionary does not.

One occasionally comes across elaborations of this usage, such as "when and only when", "exactly when", "precisely when" and so on, all apparently meaning "if and only if".

The usage "if whenever" evidently is motivated by the desire to avoid two if's in a row, for example in the sentence, "A relation $\alpha$ is symmetric if whenever $x \, \alpha \, y$ then $y \, \alpha \, x$".

See **time**.

*Citations:* (41), (98), (160), (165), (299), (338), (405), (413).

*Reference:* This discussion follows [Bagchi and Wells, 1998a].

**where**   **Where** is used in two special ways in the **mathematical register**.

### (a) To state a postcondition

**Example 1**   "**Definition:** An element $a$ of a group is **involutive** if $a^2 = e$, where $e$ is the identity element of the group." Here the statement "where $e$ is the identity element of the group" is a **postcondition**. *Citations:* (26), (275), (354).

**Remark 1**   [Krantz, 1997], page 44 and [Steenrod *et al.*, 1975], page 38, both **deprecate** this usage.

### (b) Used to introduce a constraint

**Example 2**   "A point $x$ where $f'(x) = 0$ is a critical point." In contrast to the first usage, I have not found **citations** where this usage doesn't carry a connotation of **location**. *Citation:* (254).

*Acknowledgments:* Michael Barr for references.

**without loss of generality**   A proof of an **assertion** involving two elements $x$ and $y$ of some **mathematical structure** $S$ might ostensibly require consideration of two cases in which $x$ and $y$ are related in different ways to each other; for example for some **predicate** $P$, $P(x, y)$ or $P(y, x)$ could hold. However, if there is a symmetry of $S$ that interchanges $x$ and

$y$, one may need to consider only one case. In that case, the proof may begin with a remark such as,

"Without loss of generality, we may assume $P(x, y)$."

*Citations:* (77), (34).

**witness**   If $P(x)$ is a predicate with just the one **variable** $x$, a particular **object** $c$ for which $P(c)$ is **true** is a **witness** to the fact that $\exists x P(x)$ is true.

**Remark 1**   The word "witness" also has several different technical meanings in particular fields of mathematics.

## WLOG   Without loss of generality.

**writing dilemma**   The **writing dilemma** is the question: Should we teach the students how to read mathematics as it is actually written, or should we reform our writing style so that the students are less likely to get confused?

Many proposals have been made for reforming the way we write. Most of them concern making distinctions that are now not always made in writing. Some of these widely spread ideas are:

a) We should not use the **defining equation** of a **function** as the name of the function, for example saying "The function $y = x^3$". More generally, we should distinguish between expressions and functions: Instead of saying "The function $x^3$", we should say "The function $x \mapsto x^3$" (or something similar).

b) We should not reverse the order of **quantifiers** compared to the way they would be ordered in **first order logic**. Thus we should not say "There is a prime between any integer $> 1$ and its double." Instead, we should say "Between any integer $> 1$ and its double there is a prime." (See **order of quantifiers**.)

c) We should not use "if" in **definitions** for "if and only if". (This is discussed at length under **if**.)

d) We should not use **parenthetic assertions**, for example "The function $f(x) = x^3$ has a derivative that is always nonnegative." They are hard to parse.

e) We should distinguish the **parentheses** used around the **argument** of a function from parentheses used for grouping. In Mathematica, one must write $f[x]$ rather than $f(x)$.

f) We should distinguish between the equal sign used in a definition and the equal sign used in an equation (see **colon equals**).

In fact, the writing dilemma is a paper tiger, since almost all the reform efforts are surely doomed. These proposals are reminiscent of various pioneers over the years who have wanted us to speak some completely regular and logical artificial language such as Esperanto. Such efforts have failed, although Esperanto speakers still form a small subculture. Smaller scale efforts such as spelling reform also fail more often than they succeed, too, although some of the changes Noah Webster pushed for succeeded. Recently, both Germany and France tried to institute spelling reforms which have met with great resistance.

Efforts to reform the **mathematical register** are mostly doomed to failure, too. There are two kinds of reform proposals: a) *Avoid* a certain usage. b) *Change* the language in some respect. Proposals for avoiding a usage may have some merit, especially for textbooks in beginning abstract mathematics courses. On the other hand, the proposals to change the language are potentially dangerous: if adopted by textbooks but not by mathematicians in general, which is the most the reformers could expect, we could turn out students who then would have to learn another dialect to read the mathematical literature. See also **private language**.

*Acknowledgments:* Susanna Epp.

**yes it's weird**  Students sometimes express discomfort at examples that seem arbitrary in some sense.

mathematical mind 154 or 184
set 227

**Example 1**  Try using the set $\{1, 3, 5, 6, 7, 9, 11\}$ in an example; you may get some question such as "Why did you put a 6 in there?"

**Example 2**  A different sort of example is a heterogenous set such as the set $\{3, \{2, 3\}, \emptyset\}$, which has both numbers and sets as elements.

**Example 3**  Thom [1992] objects to the use of or between adjectives when the qualities are heterogenous. Thus for him "Find all the balls that are red or white" is acceptable, but not "Find all the balls that are red or large". He was discussing the use of such examples with children in school. I have not had a student express discomfort or confusion at such usage; this may be because they have been brainwashed/educated (take your pick) by the American school system.

**Remark 1**  In teaching abstract mathematics I have adopted the practice of explicitly recognizing the students' discomfort in situations such as in Examples 1 and 2 ("yes, it's weird"). I generally say something such as: allowing such constructions is necessary to do abstract mathematics. As far as I can tell this satisfies nearly everyone. I have no basis for doing this from the mathematical education literature, but it appears to me that the discomfort is real and may very well contribute to the common attitude expressed by the phrase "I don't have a **mathematical mind**".

When a teacher takes the point of view that the student should have known that such arbitrary constructions were legitimate, or otherwise engages in put-down behavior, it can only contribute to the student's feeling of not being suited for mathematics.

**you don't know shriek**  This is the indignant shriek that begins, "You mean you don't know ... !?" (Or "never heard of ... ") This is often directed at young college students who may be very bright but who simply have not lived long enough to pick up all the information a

# you don't know shriek

middle aged college professor has. I remember emitting this shriek when I discovered as a young teacher that about half my freshman calculus students didn't know what a lathe is. In my fifties the shriek was emitted at me when two of my colleagues discovered that I had never heard of the prestigious private liberal arts college they sent their offspring to.

This phenomenon should be distinguished from the annoyance expressed at someone who isn't paying attention to what is happening or to what someone is saying.

The name is mine. However, this phenomenon needs a more insulting name guaranteed to embarrass anyone who thinks of using it.

## Z

The specification language Z was invented in Britain. Some American computer scientists call it "zed" as a result, although they say "zee" when referring to the letter of the alphabet.

***The letter*** The letter Z is pronounced "zee" in the USA and "zed" in the United Kingdom and in much of the ex-British Empire.

***The integers*** The symbol $\mathbb{Z}$ usually denotes the **set** of all integers. Some authors use $\mathbb{I}$. *Citations:* (104).

Some authors and editors strongly object to the use of the blackboard bold type style exemplified by $\mathbb{R}$ and $\mathbb{Z}$.

**zero** The number zero is an integer. It is the number of elements in the **empty set**. In American college usage it is neither **positive** nor **negative**, but some college students show confusion about this.

See also **empty set** and **root**.

The **metaphors** involved with zero are discussed in [Lakoff and Núñez, 2000], pages 64ff.

# Bibliography

**Adámek, J. and J. Rosický (1994)**, *Locally Presentable and Accessible Categories.* Cambridge University Press. (152)

**Albert E. Babbitt, J. (1962)**, 'Finitely generated pathological extensions of difference fields'. *Transactions of the American Mathematical Society*, volume **102**, pages 63–81. (92)

**Arias-De-Reyna, J. (1990)**, 'On the theorem of Frullani'. *Proceedings of the American Mathematical Society*, volume **109**, pages 165–175. (92)

**Arnold, V., M. Atiyah, P. Lax, and B. Mazur, editors (2000)**, *Mathematics, Frontiers and Perspectives.* American Mathematical Society. (162)

**Asiala, M., A. Brown, D. J. DeVries, E. Dubinsky, D. Mathews, and K. Thomas (1996)**, 'A framework for research and curriculum development in undergraduate mathematics education'. In [Kaput, Schoenfeld and Dubinsky, 1996], pages 1–32. (18)

**Azzouni, J. (1994)**, *Metaphysical Myths, Mathematical Practice.* Cambridge University Press. (157)

**Bagchi, A. and C. Wells (1998a)**, 'On the communication of mathematical reasoning'. *PRIMUS*, volume **8**, pages 15–27. Internet link to this publication given on the Handbook website. (43, 92, 100, 139, 143, 159, 159, 166, 171, 175, 207, 258, 276)

**Bagchi, A. and C. Wells (1998b)**, 'Varieties of mathematical prose'. *PRIMUS*, volume **8**, pages 116–136. Internet link to this publication given on the Handbook website. (37, 43, 123, 172, 186, 232)

**Barr, M. and C. Wells (1999)**, *Category Theory for Computing Science, 3rd Edition.* Les Publications CRM, Montréal. (81)

**Bartle, R. C. (1996)**, 'Return to the Riemann integral'. *American Mathematical Monthly*, volume **103**, pages 625–632. (194, 194)

**Bateman, P. T. and H. G. Diamond (1996)**, 'A hundred years as prime numbers'. *American Mathematical Monthly*, volume **103**, pages 729–741. (215)

**Beccari, C. (1997)**, 'Typesetting mathematics for science and technology according to ISO 31/XI'. *TUGboat*, volume **18**, pages 39–48. (121, 273)

**Bellack, A. A. and H. M. Kliebard, editors (1977)**, *Curriculum and Evaluation.* McCutchan. (290)

**Benaceraff, P. (1965)**, 'What numbers could not be'. *Philosophical Review*, volume **74**, pages 47–73. (147)

**Bills, L. and D. Tall (1998)**, 'Operable definitions in advanced mathematics: The case of the least upper bound'. In [Olivier and Newstead, 1998], pages 104–111. Internet link to this publication given on the Handbook website. (70)

**Birkhoff, G. and S. Mac Lane (1977)**, *A Survey of Modern Algebra, 4th Edition.* Macmillan. (13)

**Brown, A. (2002)**, 'Patterns of thought and prime factorization'. In [Campbell and Zazkis, 2002], pages 131–138. (28)

**Brown, L. and A. Dobson (1996)**, 'Using dissonance – finding the grit in the oyster'. In [Claxton, 1996], pages 212–227. (37)

**Bruyr, D. L. (1970)**, 'Some comments about definitions'. *Mathematics Magazine*, volume **43**, pages 57–64. (123)

**Bullock, J. O. (1994)**, 'Literacy in the language of

mathematics'. *American Mathematical Monthly*, volume **101**, pages 735–743. (133, 165)

Cajori, F. (1923), 'The history of notations of the calculus'. *The Annals of Mathematics, 2nd Ser.*, volume **25**, pages 1–46. (136)

Campbell, S. R. and R. Zazkis, editors (2002), *Learning and Teaching Number Theory*. Ablex Publishing. (281, 283, 291)

Carkenord, D. and J. Bullington (1993), 'Bringing cognitive dissonance to the classroom'. *Teaching of Psychology*, volume **20**, pages 41–43. (37)

Carlson, M. P. (1998), 'A cross-sectional investigation of the development of the function concept'. In [Schoenfeld, Kaput and Dubinsky, 1998], pages 114–162. (110, 181)

Casacuberta, C. and M. Castellet, editors (1992), *Mathematical Research Today and Tomorrow*. Springer-Verlag. (290)

Chierchia, G. (1995), *Dynamics of Meaning*. The University of Chicago Press. (61)

Chierchia, G. and S. McConnell-Ginet (1990), *Meaning and Grammar*. The MIT Press. (33, 49, 53, 94, 211, 226, 233, 263)

Chin, A. E.-T. and D. Tall (2001), 'Developing formal concepts over time'. In [van den Heuvel-Panhuizen, 2001], pages 241–248. Internet link to this publication given on the Handbook website. (86, 217)

Church, A. (1942), 'Differentials'. *American Mathematical Monthly*, volume **49**, pages 389–392. (31, 104)

Claxton, G., editor (1996), *Liberating the Learner: Lessons for Professional Development in Education*. Routledge. (281)

Cobb, P., E. Yackel, and K. McClain (2000), *Symbolizing and Communicating in Mathematics Classrooms: Perspectives on Discourse, Tools, and In-*structional Design. Lawrence Erlbaum Assoc. (289)

Cornu, B. (1992), 'Limits'. In [Tall, 1992a], pages 153–166. (145)

Coulthard, M. (1994), *Advances in Written Text Analysis*. Routledge. (287)

Coxeter, H. S. M. (1980), *Introduction to Geometry*. Wiley. (200)

Dahlberg, R. P. and D. L. Housman (1997), 'Facilitating learning events through example generation'. *Educational Studies in Mathematics*, volume **33**(3), pages 283–299. (92)

Davis, P. J. (1983), 'The thread'. *The Two-Year College Mathematics Journal*, volume **14**, pages 98–104. (173)

de Bruijn, N. G. (1994), 'The mathematical vernacular, a language for mathematics with typed sets'. In *Selected Papers on Automath*, Nederpelt, R. P., J. H. Geuvers, and R. C. de Vrijer, editors, volume 133 of *Studies in Logic and the Foundations of Mathematics*, pages 865 – 935. Elsevier. (53, 158, 158, 159, 241, 244)

Dennett, D. (1991), *Consciousness Explained*. Little, Brown and Company. (162)

DeVries, D. J. (1997), 'RUMEC APOS theory glossary'. Internet link to this publication given on the Handbook website. (17, 18)

Dieudonné, J. A. (1992), *Mathematics, the Music of Reason*. Springer-Verlag. (162)

Dreyfus, T. (1992), 'Advanced mathematical thinking processes'. In [Tall, 1992a], pages 25–41. (7, 162, 213)

Dreyfus, T. (1999), 'Why Johnny can't prove'. *Educational Studies in Mathematics*, volume **38**, pages 85–109. (207)

Dubinsky, E. (1997). 'Putting constructivism to work: Bridging the gap between research and collegiate teaching practice'. Talk given at the Re-

# Bibliography

search Conference in Collegiate Mathematics Education, Central Michigan University, September 4–7, 1997. (187, 247)

**Dubinsky, E. (1999)**, 'Mathematical reasoning: Analogies, metaphors and images'. *Notices of the American Mathematical Society*, volume **46**, pages 555–559. Review of [English, 1997]. (283, 286, 288, 288, 289, 291)

**Dubinsky, E. and G. Harel (1992)**, 'The nature of the process conception of function'. In [Harel and Dubinsky, 1992], pages 85–106. (110, 181)

**Dubinsky, E., A. Schoenfeld, and J. Kaput, editors (1994)**, *Research in Collegiate Mathematics Education. I*, volume 4 of *CBMS Issues in Mathematics Education*. American Mathematical Society. (290)

**Ebbinghaus, H.-D., J. Flum, and W. Thomas (1984)**, *Mathematical Logic*. Springer-Verlag. (81, 100, 101, 152, 154)

**Edwards, C. H. and D. E. Penney (1998)**, *Calculus with Analytic Geometry, Fifth Edition*. Prentice-Hall. (254)

**Eisenberg, T. (1992)**, 'Functions and associated learning difficulties'. In [Tall, 1992a], pages 140–152. (110)

**English, L. D. (1997)**, *Mathematical Reasoning: Analogies, Metaphors and Images*. Lawrence Erlbaum Associates. Reviewed in [Dubinsky, 1999]. (165, 283, 286, 288, 288, 289, 291)

**Epp, S. S. (1995)**, *Discrete Mathematics with Applications, 2nd Ed.* Brooks/Cole. (49, 210, 213)

**Epp, S. S. (1997)**, 'Logic and discrete mathematics in the schools'. In [Rosenstein, Franzblau and Roberts, 1997], pages 75–83. (154)

**Epp, S. S. (1998)**, 'A unified framework for proof and disproof'. *The Mathematics Teacher*, volume **91**, pages 708–713. (207)

**Epp, S. S. (1999)**, 'The language of quantification in mathematics instruction'. In [Stiff, 1999], pages 188–197. (158, 263)

**Ernest, P. (1998)**, *Social Constructivism as a Philosophy of Mathematics*. State University of New York Press. Reviewed in [Gold, 1999]. (52, 284)

**Exner, G. R. (2000)**, *Inside Calculus*. Springer-Verlag. (56, 263)

**Fauconnier, G. (1997)**, *Mappings in Thought and Language*. Cambridge University Press. (46, 47)

**Fauconnier, G. and M. Turner (2002)**, *The Way We Think*. Basic Books. (46)

**Fearnley-Sander, D. (1982)**, 'Hermann Grassmann and the prehistory of universal algebra'. *American Mathematical Monthly*, volume **89**, pages 161–166. (241)

**Ferrari, P. L. (2002)**, 'Understanding elementary number theory at the undergraduate level: A semiotic approach'. In [Campbell and Zazkis, 2002], pages 97–116. (28, 158)

**Festinger, L. (1957)**, *A Theory of Cognitive Dissonance*. Stanford University Press. (36)

**Fiengo, R. and R. May (1996)**, 'Anaphora and identity'. In [Lappin, 1997], pages 117–144. (61)

**Fletcher, P. and C. W. Patty (1988)**, *Foundations of Higher Mathematics*. PWS-Kent Publishing Company. (97)

**Fourman, M. (1977)**, 'The logic of topoi'. In *Handbook of Mathematical Logic*, Barwise, J. et al., editors. North-Holland. (271)

**Fourman, M. and S. Vickers (1986)**, 'Theories as categories'. In *Category Theory and Computer Programming*, Pitt, D. et al., editors, volume 240 of *Lecture Notes in Computer Science*, pages 434–448. Springer-Verlag. (271)

**Fraleigh, J. B. (1982)**, *A First Course in Abstract Algebra*. Addison-Wes-ley. (78)

# Bibliography

**Frauenthal, J. C. and T. L. Saaty (1979)**, 'Foresight-insight-hindsight'. *The Two-Year College Mathematics Journal*, volume **10**, pages 245–254. (133)

**Fulda, J. S. (1989)**, 'Material implication revisited'. *American Mathematical Monthly*, volume **96**, pages 247–250. (49)

**Gil, D. (1992)**, 'Scopal quantifiers; some universals of lexical effability'. In [Kefer and van der Auwera, 1992], pages 303–345. (94)

**Gillman, L. (1987)**, *Writing Mathematics Well*. Mathematical Association of America. (123, 128, 195, 245)

**Gold, B. (1999)**, 'Review'. *American Mathematical Monthly*, volume **106**, pages 373–380. Review of [Ernest, 1998] and [Hersh, 1997b]. (52, 283, 285)

**Gopen, G. D. and D. A. Smith (1990)**, 'What's an assignment like you doing in a course like this?: Writing to learn mathematics'. *The College Mathematics Journal*, volume **21**, pages 2–19. (43)

**Grassman, W. K. and J.-P. Tremblay (1996)**, *Logic and Discrete Mathematics: A Computer Science Perspective*. Prentice-Hall. (99, 184, 213, 262)

**Gray, E. and D. O. Tall (1994)**, 'Duality, ambiguity, and flexibility'. *Journal for Research in Mathematics Education*, volume **25**, pages 116–140. Internet link to this publication given on the Handbook website. (18, 181, 181, 226)

**Greenbaum, S. (1996)**, *The Oxford English Grammar*. Oxford University Press. (178)

**Gries, D. and F. B. Schneider (1993)**, *A Logical Approach to Discrete Mathematics*. Springer-Verlag. (44, 87, 184, 228)

**Grimaldi, R. P. (1999)**, *Discrete and Combinatorial Mathematics, An Applied Introduction, Fourth Edition*. Addison-Wes-ley. (87)

**Guenther, R. K. (1998)**, *Human Cognition*. Prentice Hall. (195)

**Halliday, M. A. K. (1994)**, *An Introduction to Functional Grammar, Second Edition*. Edward Arnold. (61)

**Halliday, M. A. K. and J. R. Martin (1993)**, *Writing Science: Literacy and Discursive Power*. University of Pittsburgh Press. (43, 43, 217)

**Hallyn, F., editor (2000)**, *Metaphor and analogy in the sciences*. Kluwer Academic Publishers. (287)

**Halmos, P. R. (1990)**, 'Has progress in mathematics slowed down?'. *American Mathematical Monthly*, volume **97**, pages 561–568. (133)

**Hanna, G. (1992)**, 'Mathematical proof'. In [Tall, 1992a], pages 54–61. (207)

**Harel, G. (1998)**, 'Two dual assertions: The first on learning and the second on teaching (or vice versa)'. *American Mathematical Monthly*, volume **105**, pages 497–507. (245)

**Harel, G. and E. Dubinsky, editors (1992)**, *The Concept of Function*, volume 25 of *MAA Notes*. Mathematical Association of America. (110, 283, 287, 289, 289)

**Harel, G. and J. Kaput (1992)**, 'Conceptual entities and symbols'. In [Tall, 1992a], pages 82–94. (162, 241, 243)

**Harley, T. A. (2001)**, *The psychology of language: from data to theory*. Psychology Press. (18, 61)

**Hartley Rogers, J. (1963)**, 'An example in mathematical logic'. *American Mathematical Monthly*, volume **70**, pages 929–945. (101)

**Hazzan, O. (1999)**, 'Reducing abstraction level when learning abstract algebra concepts'. *Educational Studies in Mathematics*, volume **40**, pages 71–90. (164)

**Henkin, L. (1961)**, 'Some remarks on infinitely long formulas'. In *Infinitistic Methods, Proceesings*. Pergamon Press. (211)

284

# Bibliography

**Hersh, R. (1997a)**, 'Math lingo vs. plain English: Double entendre'. *American Mathematical Monthly*, volume **104**, pages 48–51. (26, 110, 128, 166, 172, 184, 224)

**Hersh, R. (1997b)**, *What is Mathematics, Really?* Oxford University Press. Reviewed in [Gold, 1999]. (52, 157, 181, 284)

**Hershkowitz, R. (1983)**, *Proceedings of the Seventh International Conference for the Psychology of Mathematics Education*. Wiezmann Institute of Science, Rehovot, Israel. (288)

**Higham, N. J. (1993)**, *Handbook of Writing for the Mathematical Sciences*. Society for Industrial and Applied Mathematics. (123)

**Hintikka, J. (1996)**, *The principles of mathematics revisited*. Cambridge University Press. (211)

**Hofstadter, D. (1995)**, *Fluid Concepts and Creative Analogies*. BasicBooks. (15)

**Hopcroft, J. E. and J. D. Ullman (1979)**, *Introduction to Automata Theory, Languages, and Computation*. Addison-Wes-ley. (237)

**Huddleston, R. and G. K. Pullum (2002)**, *The Cambridge Grammar of the English Language*. Cambridge University Press. (88)

**ISO (1982)**, *Mathematical Signs and Symbols for Use in Physical Sciences and Technology, 2nd ed., ISO 31/11*, volume N.2 of *ISO Standards Handbook*. International Standards Organization. (121, 273)

**Jackendoff, R., P. Bloom, and K. Wynn, editors (1999)**, *Language, Logic and Concepts*. The MIT Press. (286, 288)

**Janvier, C., editor (1987)**, *Problems of Representation in the Teaching and Learning of Mathematics*. Lawrence Erlbaum Associates, Inc. (218)

**Jensen, K. and N. Wirth (1985)**, *Pascal User Manual and Report, 3rd ed.* Springer-Verlag. (100)

**Kamp, H. and U. Reyle (1993)**, *From Discourse to Logic, Parts I and II*. Studies in Linguistics and Philosophy. Kluwer Academic Publishers. (16, 16, 37, 49, 61, 65, 74, 76, 94, 128, 129, 130, 233, 253)

**Kaput, J., A. H. Schoenfeld, and E. Dubinsky, editors (1996)**, *Research in Collegiate Mathematics Education. II*, volume 6 of *CBMS Issues in Mathematics Education*. American Mathematical Society. (281)

**Katz, A. N., C. Cacciari, R. W. Gibbs, Jr, and M. Turner (1998)**, *Figurative Language and Thought*. Oxford University Press. (46)

**Keenan, E. L. and D. Westerståhl (1997)**, 'Generalized quantifiers in linguistics and logic'. In [van Benthem and ter Meulen, 1997], pages 837–896. (263)

**Kefer, M. and J. van der Auwera, editors (1992)**, *Meaning and Grammar: Cross-Linguistic Perspectives*. Mouton de Gruyter. (284)

**Kenschaft, P. C. (1997)**, *Math Power: How to Help your Child Love Math, Even if You Don't*. Addison-Wes-ley. (154, 213)

**Kieran, C. (1990)**, 'Cognitive processes involved in learning school algebra'. In [Nesher and Kilpatrick, 1990], pages 96–112. (162)

**Kirshner, D., editor (1998)**, *Proceedings of the Annual Meeting of the International Group for the Psychology of Mathematics Education – North America, Plenary Sessions Vol. 1*. Louisiana State University. (290)

**Knuth, D. E. (1986)**, *The TeX Book*. Addison-Wesley. (82)

**Knuth, D. E., T. Larrabee, and P. M. Roberts (1989)**, *Mathematical Writing*, volume 14 of *MAA Notes*. Mathematical Association of America. (258)

**Koenig, J.-P., editor (1998)**, *Discourse and Cognition*. CSLI Publications. (286)

**Kohl, J. R. (1995)**, 'An overview of English article

# Bibliography

usage for speakers of English as a second language'. Internet link to this publication given on the Handbook website. (20)

**Kolman, B., R. C. Busby, and S. Ross (1996)**, *Discrete Mathematical Structures, 3rd Edition*. Prentice-Hall. (197)

**Krantz, S. G. (1995)**, *The Elements of Advanced Mathematics*. CRC Press. (213)

**Krantz, S. G. (1997)**, *A Primer of Mathematical Writing*. American Mathematical Society. (13, 73, 119, 123, 195, 208, 276)

**Lakatos, I. (1976)**, *Proofs and Refutations*. Cambridge University Press. (162)

**Lakoff, G. (1986)**, *Women, Fire, and Dangerous Things*. The University of Chicago Press. (42, 46, 161, 162, 212)

**Lakoff, G. and R. E. Núñez (1997)**, 'The metaphorical structure of mathematics: Sketching out cognitive foundations for a mind-based mathematics'. In [English, 1997], pages 21–92. Reviewed in [Dubinsky, 1999]. (46, 46, 109, 147, 162, 162, 165, 228)

**Lakoff, G. and R. E. Núñez (1998)**, 'Conceptual metaphors in mathematics'. In [Koenig, 1998], pages 219–238. (162, 165)

**Lakoff, G. and R. E. Núñez (2000)**, *Where Mathematics Comes From*. Basic Books. (34, 46, 70, 82, 131, 154, 162, 164, 165, 215, 230, 259, 270, 280)

**Lambek, J. and P. Scott (1986)**, *Introduction to Higher Order Categorical Logic*, volume 7 of *Cambridge Studies in Advanced Mathematics*. Cambridge University Press. (271)

**Lamport, L. (1992)**. 'Types considered harmful'. DEC SRC Internal note. (258)

**Lamport, L. (1994)**, *LATEX User's Guide and Reference Manual, 2nd Ed.* Addison-Wes-ley. (195)

**Landau, S. I. (1989)**, *Dictionaries: The Art and Craft of Lexicography*. Cambridge University Press. (vi)

**Lanham, R. A. (1991)**, *A Handlist of Rhetorical Terms*. The University of California Press, Second Edition. (83, 165)

**Lappin, S. (1997)**, *The Handbook of Contemporary Semantic Theory*. Blackwell Publishers. (283, 288)

**Lawvere, F. W. and S. Schanuel (1997)**, *Conceptual Mathematics : A First Introduction to Categories*. Cambridge University Press. (231)

**Lawvere, W. and R. Rosebrugh (2003)**, *Sets for Mathematics*. Cambridge University Press. (231)

**Lewis, H. R. and C. H. Papadimitriou (1998)**, *Elements of the Theory of Computation, 2nd Edition*. Prentice-Hall. (101, 146)

**Lin, F. L., editor (2002)**, *2002 International Conference on Mathematics: Understanding Proving and Proving to Understand*. National Taiwan Normal University. (290)

**Lønning, J. T. (1997)**, 'Plurals and collectivity'. In [van Benthem and ter Meulen, 1997], pages 1009–1054. (37)

**Lum, L., editor (1993)**, *Proceedings of the Fourth International Conference on College Mathematics Teaching*. Addison-Wes-ley. (290)

**Mac Lane, S. (1981)**, 'Mathematical models: A sketch for the philosophy of mathematics'. *American Mathematical Monthly*, volume , pages 462–472. (165)

**Mac Lane, S. and G. Birkhoff (1993)**, *Algebra, third edition*. Chelsea. (108, 109)

**Makkai, M. (1999)**, 'On structuralism in mathematics'. In [Jackendoff, Bloom and Wynn, 1999], pages 43–66. (147)

**Makkai, M. and R. Paré (1990)**, *Accessible Categories: the Foundations of Categorical Model Theory*, volume 104 of *Contemporary Mathematics*. American Mathematical Society. (152)

**Makkai, M. and G. Reyes (1977)**, *First Order Cat-*

# Bibliography

*egorical Logic*, volume 611 of *Lecture Notes in Mathematics*. Springer-Verlag. (152, 271)

**Mallows, C. L. (1991)**, 'Conway's challenge sequence'. *American Mathematical Monthly*, volume **98**, pages 5–20. (275)

**Matz, M. (1982)**, 'Towards a process model for high school algebra errors'. In [Sleeman and Brown, 1982], pages 25–50. (96)

**Maurer, S. B. (1991)**, 'Advice for undergraduates on special aspects of writing mathematics'. *Primus*, volume **1**, pages 9–30. (229)

**McCarthy, M. (1994)**, 'It, this and that'. In [Coulthard, 1994], pages 266–275. (61)

**McCawley, J. D. (1988a)**, *The Syntactic Phenomena of English, Vol. I*. The University of Chicago Press. (247)

**McCawley, J. D. (1988b)**, *The Syntactic Phenomena of English, Vol. II*. The University of Chicago Press. (247)

**McCawley, J. D. (1993)**, *Everything that Linguists have Always Wanted to Know about Logic but were Ashamed to Ask*. The University of Chicago Press. (49, 57, 183)

**McCleary, S. H. (1985)**, 'Free lattice-ordered groups represented as *o*-2 transitive *l*-permutation groups'. *Transactions of the American Mathematical Society*, volume **290**, pages 69–79. (92)

**Meel, D. E. (1998)**, 'Honors students' calculus understandings: Comparing Calculus&Mathematica and traditional calculus students'. In [Schoenfeld, Kaput and Dubinsky, 1998], pages 163–215. (162)

**Mendelson, E. (1987)**, *Introduction to Mathematical Logic*. Wadsworth and Brooks/Cole. (101, 154)

**Michener, E. (1978)**, 'Understanding understanding mathematics'. *Cognitive Science*, volume **2**, pages 361–383. (92)

**Miller, J. (2003)**, 'Ambiguously defined mathematical

terms at the high school level'. Website. Link given in the Handbook website. (77)

**Muskens, R., J. van Benthem, and A. Visser (1997)**, 'Dynamics'. In [van Benthem and ter Meulen, 1997], pages 587–648. (53)

**Nardi, E. (1998)**, 'The novice mathematician's encounter with formal mathematical reasoning'. *Teaching and Learning Undergraduate Mathematics: Newsletter*, volume **9**. Internet link to this publication given on the Handbook website. (207)

**Neidinger, R. D. and R. J. Annen III (1996)**, 'The road to chaos is filled with polynomial curves'. *American Mathematical Monthly*, volume **103**, pages 640–653. (194, 263)

**Nesher, P. and J. Kilpatrick, editors (1990)**, *Mathematics and Cognition*. ICMI Study Series. Cambridge University Press. (27, 285)

**Neufeldt, V., editor (1988)**, *Webster's New World Dictionary, Third College Edition*. Simon and Schuster. (275)

**Norman, A. (1992)**, 'Teachers' mathematical knowledge of the concept of function'. In [Harel and Dubinsky, 1992], pages 215–232. (107, 110)

**Núñez, R. (2000)**, 'Conceptual metaphor and the embodied mind: What makes mathematics possible?'. In [Hallyn, 2000]. (165)

**Núñez, R., L. Edwards, and J. Matos (1999)**, 'Embodied cognition as grounding for situatedness and context in mathematics education'. *Educational Studies in Mathematics*, volume **39**, pages 45–65. (56, 165)

**Núñez, R. and G. Lakoff (1998)**, 'What did Weierstrass really define? The cognitive structure of natural and $\epsilon - \delta$ (epsilon-delta) continuity.'. *Mathematical Cognition*, volume **4**, pages 85–101. (56, 165)

**Olivier, A. and K. Newstead, editors (1998)**, *22nd Annual Conference of the International Group for*

# Bibliography

the Psychology of Mathematics Education (PME 22), Stellenbosch, South Africa. University of Stellenbosch, South Africa. (281)

**Olson, D. (1998)**. 'When reversible steps aren't elbisreveR'. Talk at Third Annual Conference on Research in Undergraduate Mathematics Education, 19 September 1998, South Bend, Indiana. (22, 149)

**Osofsky, B. L. (1994)**, 'Noether Lasker primary decomposition revisited'. *American Mathematical Monthly*, volume **101**, pages 759–768. (221)

**Partee, B. H. (1996)**, 'Formal semantics in linguistics'. In [Lappin, 1997], pages 11–38. (226)

**Piere, S. E. B. and T. E. Kieren (1989)**, 'A recursive theory of mathematical understanding'. *For the Learning of Mathematics*, volume **9**, pages 7–11. (162, 180)

**Pimm, D. (1983)**, 'Against generalization: Mathematics, students and ulterior motives'. In [Hershkowitz, 1983], pages 52–56. (145)

**Pimm, D. (1987)**, *Speaking Mathematically*. Routledge and K. Paul. (156, 158, 224)

**Pimm, D. (1988)**, 'Mathematical metaphor'. *For the Learning of Mathematics*, volume **8**, pages 30–34. (158, 165)

**Pinker, S. and A. Prince (1999)**, 'The nature of human concepts: Evidence from an unusual source'. In [Jackendoff, Bloom and Wynn, 1999], pages 221–262. (42, 43, 198)

**Pomerance, C. (1996)**, 'A tale of two sieves'. *Notices of the American Mathematical Society*, volume **43**, pages 1473–1485. (11)

**Presmeg, N. C. (1997a)**, 'Generalization using imagery in mathematics'. In [English, 1997], pages 299–312. Reviewed in [Dubinsky, 1999]. (162)

**Presmeg, N. C. (1997b)**, 'Reasoning with metaphors and metonymies in mathematics learning'. In [English, 1997], pages 267–280. Reviewed in [Dubinsky, 1999]. (46, 246)

**Raymond, E. (1991)**, *The New Hacker's Dictionary*. Massachusetts Institute of Technology. (3, 5, 138)

**Resnick, L. B. (1987)**, *Education and Learning to Think*. National Academy Press. (224)

**Resnick, L. B., E. Cauzinille-Marmeche, and J. Mathieu (1987)**, 'Understanding algebra'. In [Sloboda and Rogers, 1987], pages 169–203. (96)

**Richter, R. B. and W. P. Wardlaw (1990)**, 'Diagonalization over commutative rings'. *American Mathematical Monthly*, volume **97**, pages 223–227. (91)

**Rosen, K. (1991)**, *Discrete Mathematics and its Applications, Second Edition*. McGraw-Hill. (87)

**Rosenstein, J. G., D. S. Franzblau, and F. S. Roberts, editors (1997)**, *Discrete Mathematics in the Schools*. American Mathematical Society and National Council of Teachers of Mathematics. (283)

**Ross, K. A. and C. R. B. Wright (1992)**, *Discrete Mathematics, 3rd Edition*. Prentice-Hall. (87, 97)

**Rota, G.-C. (1996)**, *Indiscrete Thoughts*. Birkhauser. (76, 80, 232)

**Rota, G.-C. (1997)**, 'The many lives of lattice theory'. *Notices of the American Mathematical Society*, volume **44**, pages 1440–1445. (85)

**Rudin, W. (1966)**, *Real and Complex Analysis*. McGraw-Hill. (215)

**Schoenfeld, A. (1985)**, *Mathematical Problem Solving*. Academic Press. (85, 243)

**Schoenfeld, A., editor (1987a)**, *Cognitive Science and Mathematics Education*. Lawrence Erlbaum Associates. (288)

**Schoenfeld, A. (1987b)**, 'What's all the fuss about metacognition?'. In [Schoenfeld, 1987a]. (224)

**Schoenfeld, A., J. Kaput, and E. Dubinsky, editors (1998)**, *Research in Collegiate Mathematics Education III*. American Mathematical Society.

# Bibliography

(282, 287)

**Schwartzman, S. (1994)**, *The Words of Mathematics.* American Mathematical Society. (5, 82, 117)

**Schweiger, F. (1994a)**, 'Die Aesthetik der mathematischen Sprache und ihre didaktische Bedeutung'. In *Genießen – Verstehen – Veränder. Kunst und Wissenschaft im Gespräch*, Kyrer, A. and W. Roscher, editors, pages 99–112. Verlag Ursula Müller-Speiser. (158)

**Schweiger, F. (1994b)**, 'Mathematics is a language'. In *Selected Lectures from the 7th International Congress on Mathematical Education*, Robitaille, D. F., D. H. Wheeler, and C. Kieran, editors. Sainte-Foy: Presses de l'Université Laval. (158)

**Schweiger, F. (1996)**, 'Die Sprache der Mathematik aus linguistischer Sicht'. *Beiträge zum Mathematikunterricht*, volume **1996**, pages 44–51. (17, 158)

**Selden, A. and J. Selden (1992)**, 'Research perspectives on conceptions of functions'. In [Harel and Dubinsky, 1992], pages 1–16. (109, 110)

**Selden, A. and J. Selden (1997)**, 'Constructivism in mathematics education—what does it mean?'. Internet link to this publication given on the Handbook website. (52)

**Selden, A. and J. Selden (1998)**, 'The role of examples in learning mathematics'. Internet link to this publication given on the Handbook website. (90, 92)

**Selden, A. and J. Selden (1999)**, 'The role of logic in the validation of mathematical proofs'. Technical Report 1999-1, Department of Mathematics, Tennessee Technological University. (143, 175, 253)

**Sfard, A. (1991)**, 'On the dual nature of mathematical operations: Reflections on processes and objects as different sides of he same coin.'. *Educational Studies in Mathematics*, volume **22**, pages 1–36. (181)

**Sfard, A. (1992)**, 'Operational origins of mathematical objects and the quandary of reification – the case of function'. In [Harel and Dubinsky, 1992], pages 59–84. (110, 181)

**Sfard, A. (1994)**, 'Reification as the birth of metaphor'. *For the Learning of Mathematics*, volume **14**, pages 44–55. (165, 259)

**Sfard, A. (1997)**, 'Commentary: On metaphorical roots of conceptual growth'. In [English, 1997], pages 339–372. Reviewed in [Dubinsky, 1999]. (165, 259)

**Sfard, A. (2000a)**, 'Steering (dis)course between metaphor and rigor: Using focal analysis to investigate the emergence of mathematical objects'. *Journal for Research in Mathematics Education*, volume **31**, pages 296–327. (157)

**Sfard, A. (2000b)**, 'Symbolizing mathematical reality into being: How mathematical discourse and mathematical objects create each other'. In [Cobb, Yackel and McClain, 2000], pages 37–98. (157)

**Silver, E., editor (1985)**, *Teaching and Learning Mathematical Problem Solving.* Erlbaum. (290)

**Simpson, S. G. (2000)**, 'Logic and mathematics'. In *The Examined Life: Readings from Western Philosophy from Plato to Kant*, Rosen, S., editor, pages 577–605. Random House Reference. (146)

**Sleeman, D. and J. Brown, editors (1982)**, *Intelligent Tutoring Systems.* Academic Press. (287)

**Sloboda, J. A. and D. Rogers, editors (1987)**, *Cognitive Processes in Mathematics.* Clarendon Press. (288)

**Solow, D. (1995)**, *The Keys to Advanced Mathematics: Recurrent Themes in Abstract Reasoning.* BookMasters Distribution Center. (255)

**Steenrod, N. E., P. R. Halmos, M. M. Schiffer, and J. A. Dieudonné (1975)**, *How to Write Mathematics.* American Mathematical Society. (52, 99, 115, 116, 119, 158, 159, 276)

# Bibliography

**Stiff, L. V., editor (1999)**, *Developing Mathematical Reasoning in Grades K-12*. NCTM Publications. (283)

**Tall, D., editor (1992a)**, *Advanced Mathematical Thinking*, volume 11 of *Mathematics Education Library*. Kluwer Academic Publishers. (113, 162, 282, 282, 283, 284, 284, 290, 290, 291)

**Tall, D. (1992b)**, 'The psychology of advanced mathematical thinking'. In [Tall, 1992a], pages 3–21. Internet link to this publication given on the Handbook website. (144, 161, 162)

**Tall, D. (1992c)**, 'Reflections'. In [Tall, 1992a], pages 251–259. Internet link to this publication given on the Handbook website. (243)

**Tall, D. (1993)**, 'Conceptual foundations of the calculus'. In [Lum, 1993], pages 73–88. (145)

**Tall, D. (1999)**, 'The cognitive development of proof: Is mathematical proof for all or for some?'. In *Developments in School Mathematics Education Around the World*, Usiskin, Z., editor, volume 4, pages 117–136. National Council of Teachers of Mathematics. Internet link to this publication given on the Handbook website. (207)

**Tall, D. (2001)**, 'Conceptual and formal infinities'. *Educational Studies in Mathematics*, volume **48**, pages 199–238. (131)

**Tall, D. (2002)**, 'Differing modes of proof and belief in mathematics'. In [Lin, 2002], pages 91–107. Internet link to this publication given on the Handbook website. (154, 207)

**Tall, D. and S. Vinner (1981)**, 'Concept image and concept definition in mathematics with particular reference to limits and continuity'. *Educational Studies in Mathematics*, volume **22**, pages 125–147. (39, 70, 145, 162)

**Thom, R. (1992)**, 'Leaving mathematics for philosophy'. In [Casacuberta and Castellet, 1992], pages 1–12. (279)

**Thompson, P. W. (1985)**, 'Experience, problem-solving, and learning mathematics: Considerations in developing mathematical curricula'. In [Silver, 1985], pages 189–236. Internet link to this publication given on the Handbook website. (7)

**Thompson, P. W. (1994)**, 'Students, functions and the undergraduate curriculum'. In [Dubinsky, Schoenfeld and Kaput, 1994], pages 21–44. Internet link to this publication given on the Handbook website. (18, 107, 110, 218)

**Thompson, P. W. and A. Sfard (1998)**, 'Problems of reification: Representations and mathematical objects'. In [Kirshner, 1998], pages 1–32. Internet link to this publication given on the Handbook website. (162, 181, 219)

**Thurston, W. P. (1990)**, 'Mathematical education'. *Notices of the American Mathematical Society*, volume **7**, pages 844–850. (213)

**Vallance, E. (1977)**, 'Hiding the hidden curriculum: An interpretation of the language of justification in nineteenth-century educational reform'. In [Bellack and Kliebard, 1977], pages 590–607. (62)

**van Benthem, J. and A. ter Meulen (1997)**, *Handbook of Logic and Language*. The MIT Press. (285, 286, 287, 290)

**van Dalen, D. (1989)**, *Logic and Structure*. Springer-Verlag. (101, 154)

**van den Heuvel-Panhuizen, M., editor (2001)**, *25th Annual Conference of the International Group for the Psychology of Mathematics Education (PME 25), Utrecht*. University of Utrecht, The Netherlands. (282)

**van Eijck, J. and H. Kamp (1997)**, 'Representing discourse in context'. In [van Benthem and ter Meulen, 1997], pages 179–239. (74)

**Vaught, R. L. (1973)**, 'Some aspects of the theory of

models'. *American Mathematical Monthly*, volume **80**, pages 3–37. (80)

**Vinner, S. (1992)**, 'The role of definitions in the teaching and learning of mathematics'. In [Tall, 1992a], pages 65–81. (43, 70)

**Vinner, S. and T. Dreyfus (1989)**, 'Images and definitions for the notion of function'. *Journal for Research in Mathematics Education*, volume **20**, pages 356–366. (39, 70, 110, 219)

**Watson, A., P. Spirou, and D. Tall (2003)**, 'The relationship between physical embodiment and mathematical symbolism: The concept of vector'. To be published in The Mediterranean Journal of Mathematics Education. (274)

**Weil, A. (1992)**, *The Apprenticeship of a Mathematician*. Birkhaeuser Verlag. Translated from the French by Jennifer Gage. (82)

**Wells, C. (1995)**, 'Communicating mathematics: Useful ideas from computer science'. *American Mathematical Monthly*, volume **102**, pages 397–408. Internet link to this publication given on the Handbook website. (3, 44, 93, 110, 162, 171, 224, 228, 232, 258, 258)

**Wells, C. (1997)**, 'Discrete mathematics'. Class notes, Case Western Reserve University. (19, 60, 228)

**Wheatley, G. H. (1997)**, 'Reasoning with images in mathematical activity'. In [English, 1997], pages 281–298. Reviewed in [Dubinsky, 1999]. (162)

**Wolfram, S. (1997)**, *The Mathematica Book*. Wolfram Media. (244)

**Wood, L. (1999)**, 'Teaching definitions in undergraduate mathematics'. Internet link to this publication given on the Handbook website. (70)

**Wood, L. and G. Perrett (1997)**, *Advanced Mathematical Discourse*. University of Technology, Sydney. (94, 263)

**Zalta, E. N., editor (2003)**, 'Stanford encyclopedia of philosophy'. Internet link to this publication given on the Handbook website. (152)

**Zazkis, R. (2002)**, 'Language of number theory: Metaphor and rigor'. In [Campbell and Zazkis, 2002], pages 83–96.

**Zulli, L. (1996)**, 'Charting the 3-sphere—an exposition for undergraduates'. *American Mathematical Monthly*, volume **103**, pages 221–229. (194)

# Index

# Index

# Index

# Index

# Index

# Index

# Index

# Index